# STARTING A STUDENT / NONCOMMERCIAL
# Radio Station

## JAMES J. MCCLUSKEY, PH.D.

SIMON & SCHUSTER CUSTOM PUBLISHING

Cover photos courtesy of ERI–Electronics Research,
7777 Gardner Rd., Chandler, IN 47610.

Prudential antenna was photographed by Paul Shulins.
Empire State Building was photographed by Lou Bopp.

Printed in the United States of America

10  9  8  7  6  5  4  3  2  1

*Please visit our web site at www.sscp.com*

ISBN 0–536–01067–6

BA 98033

SIMON & SCHUSTER CUSTOM PUBLISHING
160 Gould Street/Needham Heights, MA 02194
Simon & Schuster Education Group

# Table of Contents ▶ ▶

# Acknowledgments ▶ ▶

I would like to express my heartfelt appreciation to the many people who have helped to make this book a reality! First and foremost, to Bridget McIntyre Troy, Editor of Simon & Schuster Custom Publishing, whose professional guidance, sense of humor, expedited e-mailing and patient understanding have made this project possible! I also owe a huge debt of gratitude to my family, including my loving wife, Mary Sue, my children, Amanda, Melissa, Adam & Christy (who never quite understood why daddy was always working at the computer), to my mother (Kay), my sister Mary Ann and my brother Jack (John), without whose moral support I would have been hopelessly lost. My heartfelt appreciation is also owed to Central Michigan University Professors Dr. Jerry Henderson (for his contribution of chapter 15 of this book), and to Dr. B.R. Smith for editing the completed manuscript. I also want to express my appreciation to the many contributing organizations and individuals who made this book possible. It is to these many wonderful people and to you that I dedicate this book.

# Chapter 1 ▶ ▶

## ESTABLISHING OBJECTIVES, GOALS, AND CHARTING YOUR COURSE

---

It has been said that anything is possible, given enough time and money. The desire and determination to see a project through to its fulfillment are also essential elements in reaching your goals. One of the biggest challenges to starting a student station is the inherent problem of turnover. Inevitably the ambitious students who had the vision to initiate the project won't be the same ones who will later be enjoying the fruits of this labor. The advantages of fresh creativity, new vision and youthful energy are offset by the likelihood that in a few years these innovators will be graduating and moving on to greater challenges elsewhere. Yet student media is shaping the media leaders of tomorrow. Still, student media do have the potential for shaping the broadcast leaders of tomorrow. It's an exciting forum for creativity, learning, and experience.

Starting a student radio station involves a great deal of thought and planning. This is a bit like "looking into a crystal ball" and trying to foresee the kinds of obstacles that must be overcome.

A primary factor for success is strong leadership and supervision. Whether a station has close faculty supervision or is "loosely managed," students still benefit by having a student radio station on their campus. Any efforts to start such a station will go much further with a strong following of students, faculty and administration at your school. How the student station is organized (from a managerial perspective) seems to have little effect on its success. However, careful and detailed planning will smooth the process and greatly enhance the likelihood of success in completing your project.

### Selecting a Faculty or Administrative Advisor

Selecting the right person to help in the effort can mean the difference between life and death, or success and failure for a new station. You will probably find that a potential advisor's shared interest in starting the radio station is at least as important as personal expertise in the field.

Hypothetically you could have a choice between a veteran radio engineer on the faculty with years of professional experience, and a youthful but inexperienced teacher who firmly believes in what your group is trying to accomplish. Many times enthusiasm and drive will get you through some very hard times. This is not to discredit experience, but rather to help you find someone who won't give up supporting your project when the going gets tough.

Depending on the circumstances, there can be a lot of politics involved in getting a new student radio station off the ground. Assuming you have a group of students who really share the dream of having a station who have desire and ambition in the group's nucleus, you must first seek a supportive, understanding faculty member or administrator who will see the group's project through to completion. There are many hurdles that must first be cleared before you sign on the air for the first time. As you take the very first steps toward achieving your goal, there is no way to foresee all of the challenges that lie ahead. The whole process combines personalities, politics, pressure, stress and desire before success can be achieved.

## Selecting Your Leaders

Although your organization or student group can take many directions, carefully choosing which student leader(s) will best represent your interests does make a big difference. You will want to select someone who is reliable, dependable, trustworthy, and who will not be afraid to stand up and fight for the group's interests if the need arises. It isn't necessary to pick the most popular person in your school. It is not a prerequisite that they were voted as most likely to succeed or best liked, either. What is important is that this offers a track record indicating the leader has successfully finished what was started in the past. They don't need to be liked by everyone in your group, but the majority of your group's members should be comfortable with the person you collectively select.

You may wish to choose a committee. But it is often the experience of groups that committees seldom get very much accomplished without some form of individual leadership. Choose someone who really wants the job and who will always genuinely represent your group's interests.

## Public Relations

Starting (and keeping) a student radio station is really a major public relations (or PR) challenge right from the beginning. It is definitely best to gain the approval of your school's administration before undertaking such an effort. You will find that having the support of a friendly faculty and administration is a strong advantage when the winds of opposition begin to blow against your efforts.

Some students may be rather rebellious and prefer to take matters into their own hands. Perhaps this is an outgrowth of the "me first" and "I want it now" sentiments of our society. In these times of instant gratification, it is understandable that some people would not be willing to wait for the procedural wheels of democracy and the government to turn.

These "pirate" efforts are further discussed later in this book.

## Building Positive Faculty Relations

Students seldom realize that while they come and go, a school's faculty and administrators are the ones who have to stay and live with the radio station. There are both administrative joys and headaches associated with such an operation. Winning over your school's faculty and administration is one of the great PR challenges that lies ahead of you. If something goes out on the air that is slanderous, or expresses an unpopular viewpoint, or is considered to be profane or indecent to some of your listeners, you must realize that it is someone in the faculty or administration who is ultimately responsible. Students may have great freedom and at some schools may even be challenged to express their opinions, but there is usually a representative of the school who will ultimately have to defend the actions of the students against such charges.

With this in mind, it is strongly advised that you approach your faculty and administration gently with the idea of starting a student radio station. Practice your public relations techniques in selling them on the concept: it is best if you approach the idea from a series of positive perspectives, explaining how much this can mean to the school, to the students, and how this will help the school in its educational mission. Be sure you consider and list in writing all of the advantages of having a student radio station on your school's campus. If you can convince administrators that the advantages of having a student radio station far outweigh any disadvantages, you will have made great progress in making your goal a reality.

## Determining Your Course of Action

Once you have established a core group of interested students dedicated to seeing the project through, and after you have received written support from your school's faculty and administration, it is time to chart a specific course of action. In this author's book titled *Broadcast Station Management and Ownership: The Professional, Personal Perspective*, the first chapter is dedicated to discussing how to write a list of goals and how to determine objectives to reach each of them. Goals should be clearly specified, committed to writing, and posted in an obvious location that will be seen daily. Only by being reminded of these goals and objectives, along with established deadlines for each, will you successfully strive to reach them. They must constantly be in your mind if your attention and actions are to be focused on them.

Once your group leaders and your school's faculty and administrative support groups have agreed on a list of deadlines for the objectives and the goals you wish to reach, you (collectively or individually) are ready for action.

## Exploring Options

Your school's needs should be identified and explored before making decisions about which actions are best. In some cases, multiple options might actually be better for your school than simply making a single choice.

Faculty members and administrators are often somewhat hesitant to commit themselves to supporting such a new venture. By nature, everyone fears the unknown, and a faculty member may not be eager to volunteer for additional responsibilities. Stories of students at other stations acting irresponsibly and causing grief for their faculty advisors may come through the rumor mill, making it even more difficult to build faculty support.

In these cases, faculty and administrators are likely to be more willing to offer their support if the students can prove their ability to act in a responsible manner. Should your group be confronted with this kind of response, you may wish to "drop back and punt" or go back to the basics.

If your request is to establish a federally-licensed, over-the-air AM or FM broadcast station, and your faculty or administration wavers in their support, it might be in your student group's best efforts to suggest an alternative plan. It may be better to have your proverbial "foot in the door" than to have the door slammed in your face altogether.

## Available Options

### Junior/Senior High School Public Address System

A quick and easy way to distribute music and programming through a junior or senior high school is to utilize the existing public address system. Such a system is often already in place and is usually used on a daily basis for school-wide announcements in the morning. An interested broadcasting club of students could approach the school principal and set up a schedule when they could possibly help with the announcements and make them a "school news" format, create a little programming to accompany the news, and further build the schedule from there. Our high school administration also permitted us to "pipe" music through the cafeteria during the school lunch periods. We also arranged a trade for records with a local business in exchange for mentioning their name (in the early days of "underwriting"), which turned out to be successful for both parties. Your school may permit this, or it may not, depending your school's administration. It's worth a try: at worst they can simply say no. If this is the response, try to find out why, and go on from there. Often things can be worked out and you may be able to begin your project on a trial basis, perhaps under more direct supervision, at least.

## Carrier Current

If your school (such as a college or university) has buildings where students reside on campus in residence halls (formerly known as dormitories), a possible option to an on-air station is carrier current broadcasting. Carrier current broadcasting utilizes the existing electrical wiring inside the walls of buildings, and therefore does not require an FCC license. This system is limited to radiating only about ten to twelve feet from an electrical circuit, which generally means it is adequate to cover the interior of most buildings.

Carrier current began as an Amplitude Modulation (AM) system, but the technology has been developed to permit Frequency Modulation (FM) and FM Stereo transmissions. Most AM carrier current broadcasters prefer the low end of the dial (around 540 MHz. or higher) which permits the best radiation distances. It is important that your group works closely with the school's plant department to maintain best working relationships and to ensure positive future cooperation from them. If you need access to equipment in the future (and you most assuredly will), friends in the plant department will prove invaluable to your station. These people may also prove to be very helpful in making suggestions and in helping to make minor installations which can prove to be more convenient and improve safety and security. Examples of these installations or modifications might include: 1) making physical changes to a room such as installing a new door or wall; moving an existing door or wall, installing or moving lighting and/or windows; 2) installing new locks or a security system in your building; or 3) installation of an intercom system; or 4) installing new telephones or moving/relocating existing telephones, and such.

In establishing your carrier current station, it is important that you tune your system to a frequency that will not interfere with any other licensed stations which can be received in your area. This is essential if you are to maintain good relations with those who already listen to established stations in your area. You never know when you may need to call on such stations for help, equipment donations, and such. This also is simply good engineering practice. Interfering with an existing broadcast signal is also in violation of the Federal Communication Commission's rules and regulations.

Making good equipment choices is also vital. First, it is important that even carrier current transmitters are of a type that is acceptable to the Federal Communications Commission (FCC type accepted). This means that the design was approved by the Commission, and must not be modified from the original intent for this equipment. With the relatively low cost of carrier current transmission equipment, it is often a good idea to at least check with several distributors of carrier current equipment. Such manufacturers as LPB, Inc. (Low Power Broadcasting) can be of great help in advising you with your carrier current system.

## Leaky Cable

A recent technological development that is a variation on the carrier current idea is called "leaky cable." By sending a radio frequency (RF) signal into a specified length of coaxial cable, a signal will be radiated. When this cable is hung down an elevator shaft or down the center of a building, the signal is usually sufficient to reach radios throughout the building, particularly if the building isn't very wide, but is tall (which is especially likely in such applications as college student residence halls).

Again, it is best to get expert advice from such professionals at companies which distribute this kind of product. Leaky cable can be superior to carrier current because it does not require that you interface the transmitter with the existing electrical system. Again, as with carrier current, leaky cable falls under Part 15 of the FCC rules, and does not require a license. However, the transmitter must be FCC type accepted, under the same rules outlined in the carrier current segment of this chapter.

## Cable FM

A community cable television operation offers another option you may wish to investigate. Many cable operators have a "public access" channel which can be an excellent forum for student-produced television programming. In addition, often cable operators have other services such as weather channels, community news channels, and university channels. While the video on these channels may be programmed by the cable operator with such devices as character generators, computer hard drives, and such, often the audio can be provided locally by students at either the junior/senior high school or college level. With the addition of a transponder (which you can negotiate with the cable operator to provide), your school can almost immediately begin sending audio programming over a cable channel with crystal-clear sound.

Again, no FCC license is required, and often cable operators are happy to have locally produced programming on their system.

With cable FM channels, there is also the opportunity for fundraising. However, any such sale of advertising or underwriting should be negotiated with your cable operator prior to beginning any programming. There is a wide variety of opportunities for the sale of advertising. While some cable operators expect a percentage of the gross revenues, others may not. This is definitely something which you should discuss and have committed to a written agreement before you commence your programming.

## Over-the-Air AM Broadcasting

Because AM suffers from "scratchy, static-filled audio" which is susceptible to noise from terrestrial sources such as fluorescent lights or electrical storms, students sometimes view this form of radio as undesirable. However, the AM band may offer the best opportunity for student broadcasting. Because of its declining popularity, often the AM band allows students to reach larger audiences, especially true in crowded metropolitan areas where finding a frequency on the FM band is nearly impossible.

There are more and more cases of AM broadcasters deciding to give up and donating their licenses to local schools. This practice assures the local broadcaster that they are not creating more external commercial competition for their co-owned FM facility.

## Over-the-Air FM Broadcasting

Although the FCC no longer issues class D licenses (i.e. ten watt and less FM licenses) for stations, there are still opportunities for local and community licenses of 100 watts and more.

There are sometimes occasions where an existing station will share time or allow local schools to have an hour of weekend or late evening airtime to produce and air their own program. Although this is less desirable than having your own 24 hour station, it is a much easier path to getting students on the air, and provides the same "taste" of producing a program. It is excellent practical experience in a "hands on" learning environment.

This author had the opportunity to produce a one hour weekly program with a small group of other students while he was in high school. The Junior Achievement program encourages learning about business by allowing students to form small companies, to sell stock to raise operating capitol, to produce and sell products, to keep accurate accounting records, and to then liquidate their company and the stock. Our Junior Achievement company sold stock, then used this capitol to buy an hour of airtime on a local radio station.

We also sold commercials during the hour of programming, and the process worked very well. Our company offered one of the best returns on investment, and we would have paid dividends of 200% if it had been permitted under Junior Achievement rules. Our pleasant dilemma was what to do with all the profits at the end of the school year. We elected to hold a company party, which was appreciated by all of the participants.

Sharing time or buying time on an existing local FM commercial station, or working out an arrangement with an existing noncommercial station is definitely an option that you and your group should consider. However, if your school's group is determined to have its own FM over-the-air station, you should be prepared for a long and rigorous process. Contending with the legalities and bureaucracy can be frustrating, particularly if you do not have previous experience dealing with the government. This is especially true if you do not have the right kinds of support groups in place. This includes having knowledgeable people available to give good advice, and deciding whether the advice is good for your group. In populated metropolitan areas, the chances of your group's winning an FM broadcasting license may be lower than gambling successfully in Las Vegas.

The bottom line is that there is a very limited number of frequencies available, and federal law has established that these frequencies belong to the people of the United States. Any person or entity holding a license or licenses is merely a caretaker of that frequency for a fixed period of time.

There is a lot of planning you should do before tackling a project like starting a new radio station. In summary, you must first determine what kind of station your group desires and how to practically go about getting organized. The process involves hiring an experienced and knowledgeable broadcast engineer or broadcast consulting firm with significant experience in dealing with matters before the FCC. After you feel confident that you, A) have excellent long-term support in your school from your fellow students, faculty, and administrators; and, B) have engaged the services of a qualified broadcast consulting firm, the next step in the process is to have your engineer provide you with available frequency options and power levels. You may find that your consultant suggests several available options, but you must allow time to carefully consider the possibilities. At first glance it may appear that the option which offers the most power from the tallest tower is what you want. However, this isn't always the option that will best serve your school's needs. Unless your school already has some kind of broadcasting station in operation, you may not be prepared for a full-power FM that can cover several counties. You may also not be prepared for the associated battle for the license that may be looming on the horizon.

If the frequency has already been allocated to your community by the FCC, you'll merely have to file an FCC form 301 (application for a commercial FM radio station construction permit) or FCC form 340 (for a noncommercial FM radio station construction permit). Typically your technical consultant will complete the engineering portion of the application form for you.

The process becomes more complicated if the frequency has not been allocated to your community. Because FM stations are so popular with listeners, the process to win a license is very competitive. Licenses in metropolitan areas are usually already taken, which leaves few options for schools near major population centers. A creative broadcast engineering consultant may find you a good frequency and power combination that well suits your needs. The next step is to petition the FCC to have this frequency allocated to your community or city. If you are in a small community, prerequisites for determining if a "village" is qualified include having a United States Post Office, a form of local self-government in place, and other similar considerations.

When you petition the FCC to have a frequency allocated, you must also include a statement which amounts to a letter of intent indicating you will apply for the frequency if it is allocated. It typically takes 60 to 90 days for the FCC's engineering branch to investigate your proposal. They must ascertain whether the frequency "fits" in your area without causing interference to other stations on the same or adjacent frequencies. Sometimes your consultant will propose moving other stations around to accommodate your new frequency. This complicates the process, and usually moving other station's frequencies must be done at your expense. This usually means you will be responsible for the costs including re-tuning the exciter, transmitter and antenna to the new frequency. Stations which have been long established on a given frequency aren't always too happy about this kind of proposal, especially if it means major changes in their location on the dial, which also may mean a major (and expensive) promotional effort to tell their listeners when and to where on the dial they are moving. These station licensees will like it even less if it means having another competitor moving into the marketplace. However, the FCC rules require that they obey, but they may file comments against the move, which will likely delay the process. In this case, sometimes it is a consideration you should examine before deciding to move other stations around the dial. This is usually done only as a last ditch effort to fit a new station into an already crowded FM dial.

Once your petition for allocation of a new FM frequency passes the FCC's engineering division, it is posted for public notice and comment. This means the FCC is asking for comments from anyone as to whether the new frequency fits into the local spectrum and best serves in the public interest.

The window for this comment period can also take 60–90 days or more: sometimes as many as 120 days. At that point the FCC will close the window for comments on the allocation and will wait an additional 30–60 days. During this time the commission's engineers review the comments they have received, and then they will make a recommendation to the legal division. The FCC's legal branch may then post a public notice that the frequency has been allocated to a particular area. Included with this announcement is an invitation for any qualified party to apply for the frequency.

At this point your broadcast engineering consultant will complete the engineering section of your application for a new construction permit, known as FCC form 301 or 340. This kind of information includes the coordinates of your tower location (given in latitude and longitude), tower height, and other specifics including antenna gain, coaxial cable specifications including line loss, and transmitter power output, type acceptance, and a host of other details. You must also disclose your intended studio site location (if different from the transmitter location) which will be used as the transmitter control point. If you intend to use a small transmitter, known as a studio-transmitter link (or STL), and/or remote pickup transmitting devices (for doing live broadcasting from locations outside your studios) you must also find a frequency or frequencies and apply for these licenses known as auxiliary services. These will be part of your main station license. You can apply for them after you receive your main station license, if you prefer.

In addition to preparation of your engineering section of FCC form 301 (or 340), you should already have your legal support group established. A Washington-based law firm should have been already selected using similar criteria to that used to select your engineering consultant. This includes having significant successful experience in previous matters before the FCC. It should be noted that the services of these attorneys is not inexpensive: costs can sometimes range from $200 or more per hour, depending on the nature of the services rendered.

Your FCC attorney will assist you in completing your FCC form 301 or 340 application, with exception of the engineering section, which is completed by your engineering consultant. It is important that you employ the services of a qualified Washington-based FCC law firm. It is not uncommon for groups that think they can complete their own application to have them returned by the FCC for minor flaws and oversights. If, for example, you forget to complete one blank or if you do not use certain predisposed language in some of your answers, your application may be deemed unacceptable by the FCC's legal department and your application returned. Mistakes of this nature can be costly and devastating to your effort, particularly if other parties have also filed for the same frequency. If your application is found to be unacceptable (after the filing window has been closed) and your application has been returned, other parties' applications may be acceptable. If there is only one other competing application filed with yours and your application is thrown out, the other party's application is granted because it may have been the only acceptable application filed with the commission for that frequency.

At this point you have lost and although you do have other options, your chances of winning and reversing an FCC decision are greatly diminished. Hiring the necessary legal representation to reverse an FCC decision to grant a construction permit is a very costly investment. Foresight should tell you that it is going to be less costly to hire the Washington-based law firm which is experienced in FCC matters and which will assure that your application is complete and acceptable the first time it is filed.

This author has had significant experience with this process on both sides of the effort, and the preference (by far) is to employ the services of a qualified Washington-based law firm experienced in dealing with these kinds of FCC-related matters. You will be actually saving money over the long term, and an added benefit is that a good attorney in Washington can hand-deliver your application and accompanying copies, have them hand-stamped by the secretary of the FCC, and carry the application through the various steps and branches to make sure it doesn't wind up sitting in a pile on the desk of someone who happens to be on vacation or official leave. There have been instances of applications which were mailed to the FCC being lost, misplaced, or just disappearing in the maze of paperwork in the commissions files and offices. It is understandable that these kinds of things can happen: you just don't want them happening to your application. Be wise, shop around for good Washington-based counsel, and be prepared to pay substantial amounts of money for services that can, in the long run, prove to be invaluable.

If you have followed the timeline in this process from petitioning the FCC for frequency allocation to the actual granting of a construction permit, you have found that it can easily take more than a year. The average processing time for a commercial FM application is usually about eighteen months with noncommercial applications typically processed in nine months. If an application is crossfiled (or "MX-ed") and there are competing parties filing for the same frequency, the process can be delayed for several years. This is a long time, and in dealing with matters before the FCC, time is money . . . your money. That is why it is imperative that your effort have complete and total backing from your school's administration: they must want this station as much or more than the students do, for successful completion of this process and the winning of a station license. It can cost a great deal of money, particularly if it is a high-power station being sought, or if a frequency needs to be allocated, or if the application is crossfiled by competing parties. It is very important that student applicants and advisors realize these things from the outset and that they are prepared, financially, legally, mentally, emotionally, and totally. Note that the circumstances vary greatly, depending on several variables which include: A) the location of the proposed new station (a rural or a metropolitan population center), B) whether it is AM or FM, C) whether it is commercial or noncommercial, D) the station's proposed coverage area (is it high-power or low-power), and sometimes E) the complexity of construction (for example, whether the antenna going to be placed on a mountain, whether it will require excessive construction costs such as a very tall tower, whether it requires substantial negotiation for land to construct the station, whether it will involve dealings with the Federal Aviation Agency (FAA) to approve tower construction, whether there will be excessive costs to run power to the transmitter side, and a host of other similar considerations which we will discuss and explore in upcoming chapters.

## Commercial vs. Noncommercial

The majority of student radio stations are noncommercial. The advantages of being noncommercial can outweigh those of being commercial, depending on your school's goals and objectives for having a station, and for the nature of service to your community. This decision should be made early in your planning process, as it can determine how much of an investment the process might involve over the long term, as well as how support for your new station will be structured.

Under de-regulation and other rules, the FCC has lifted and will continue to ease the former, tight regulation of what announcements constitute underwriting vs. commercials. Still, it is illegal to air commercials on a noncommercial station. It could be misleading to list here the current regulations concerning what is permitted and what is disallowed as content in underwriting announcements because the rules seem to be changing almost on a daily basis. Because of the importance of this matter, however, it is essential that the station's policy on this be current and clearly understood by all staff members.

It should be noted that commercial station licenses are more sought-after commodities than their noncommercial counterparts. Your school's goal of having an over-the-air broadcast station may be more easily fulfilled if you decide to seek a noncommercial license. Of course, this statement is a generalization, and your own specific circumstances and variables must be explored to better make this decision. Schools in rural areas, for example, may have more frequency options and less competition for frequencies. Therefore, they may have a better chance at being granted a commercial broadcast construction permit. It is possible to change a license from commercial to noncom-

# Chapter 2 ▶ ▶

## STARTING A LICENSED STUDENT RADIO STATION

### Outline: Finding a Frequency for Over-the-Air AM or FM

A.  Frequency search (how its different for AM vs. FM)

B.  Allocation of frequency process

C.  FCC public notice process

D.  Cross filing potential from competing applicants

E.  The minority cross filing possibility

F.  Cutting costs for the frequency search (what it might cost to try to cut costs)

　　1.Ways to do it yourself using FCC database or engineering students at your school

G.  Classes of Noncommercial Educational FM stations and channels

## Part One: Frequency Searches ▶

Congratulations! Your school has decided to accept a brand new challenge: you are starting an over-the-air student radio station. At this time it is important to determine whether you want to broadcast in Amplitude Modulation (AM) or Frequency Modulation (FM). Because of the extreme popularity of FM Stereo broadcasting, you may think it is silly these days to even consider starting a new AM station. However, it is just this popularity that can sometimes make obtaining a new FM construction permit more difficult.

Frequency searches are best accomplished by a professional broadcast engineering firm whose credentials are a matter of record with the Federal Communications Commission. Many students are under the false impression that they can go out to their car radio, tune up and down the dial, find a blank spot where nothing but static can be heard, and apply for that frequency. Unless you or someone you know has experience in this area, you are best advised to hire a professional engineer. Some additional advice is to copyright the engineering section of your application to protect it against possible theft by other potential competing applicants. This kind of theft has recently surfaced in the arena of applications for new station construction permits on the FM band.

## AM Frequency Searches

Finding an AM frequency is almost like "discovering" a blank spot on the dial. It is also a matter of first adjacencies and second adjacencies, channel separations, mileage separations, signal contours, and (if fulltime) nighttime skywave interference predictions. If this sounds like "Greek" to you, plan on hiring someone who knows how to do it.

Sometimes, particularly near crowded metropolitan areas, you may actually have a reasonable chance of finding a "good" frequency by installing a directional array. This may require more elaborate engineering surveys and more up front costs (not to mention additional construction costs).

Amplitude Modulation has certain unique physical propagation properties. A simplified way to identify the elements of a "good" frequency would include dial position and power level. Generally, the lower on the AM dial you are, the less power it takes to cover the same area. For example, an AM nondirectional station on 560 kilohertz (kHz.) with 1000 watts could cover more area than a station with 5,000 watts on a frequency of 1600 kHz. Some engineers have said they have known stations with 50,000 watts on the "high" end of the dial that have not covered as much as 5,000 watt stations on the "low" end of the dial.

The author's own station (a 5,000 watt AM station on 1510 kHz.) could not cover the same area as a 1,000 watt station on 970 kHz. If you are thinking about starting a new student AM station, it would be best to try to find a frequency on the low end of the dial. Most competent professional consulting engineers have software programs that will "automatically" seek the first possible/available frequency starting from the low end of the dial.

However, in this business you should never "assume" anything. Specific questions can help ensure that the choice(s) consultants may give you are the best possible ones.

If you are thinking about going "carrier current," picking a low dial position can also improve your station's reach and ability to overcome interference. With carrier current, you can choose your own frequency. The only requirement is that your station doesn't interfere with a station already on the air at or near that particular frequency. An important aspect of carrier current is that it does not require a license from the FCC, as it operates under part 15 of the commission's rules.

In terms of over-the-air AM frequencies, your consulting engineer will typically give you several different options: combinations of power levels and frequencies. Sometimes these choices will specify a directional pattern, which substantially increases the construction costs because of the need for extra tower(s), more transmission line, additional copper for the installation of ground planes radiating out from each tower, and phasing equipment. Directional stations require much more maintenance at keeping their "pattern" within established parameters.

If you are given a choice and your project is on a "limited" budget, try to steer away from directional installations. A nationally recognized consultant once said that "what you are told can be done with six towers can often be accomplished with five." In other words, the extra amount of money you may invest in hiring a very good engineering firm can actually save you substantial amounts over the long term of constructing and operating the station. Sadly there are numerous six tower directional AM arrays that could have been built more economically if proper design techniques had been utilized in their planning.

Your goal should be to seek a construction permit for a single tower full-time AM station built on the lowest frequency available which affords you the most power. In this way you will be assured of the maximum coverage area with the minimum invested. This, of course, will depend a lot on your proximity to a large metropolitan population center, as well as many other variables (like how crowded the spectrum is in your area and how well your professional consulting engineer has done his/her homework).

The process of applying for a new amplitude modulation radio station basically involves convincing the FCC's engineering bureau that there is spectrum space available and that it will not interfere with existing facilities. Usually your consulting engineering firm can use existing propagation data or computer estimates of contours from stations on adjacent frequencies to demonstrate noninterference from the proposed facility. On occasion, particularly with proposed directional stations, engineers may have to take field measurements on existing co-channels first and second adjacent frequency stations to determine their actual radiation pattern and to demonstrate through the use of the measured data that the proposed facility will not interfere with any existing facility.

It should be noted that the process of applying for a directional facility can mean a much greater investment overall. This includes a substantial amount of engineering time, which can transmit to a lot of money. In the author's experience, the cost averaged nearly $7500. If you are not prepared to make this kind of investment, you may want to investigate alternatives, such as going with less expensive options (as we have previously discussed in the first chapter).

This is primarily what is involved in the frequency search and selection process for a new AM station. The actual process of filing the application, including this engineering portion, will be discussed later.

## FM Frequency Searches

Since the implementation of docket 80–90, the Federal Communications Commission has opened up a significant number of new FM channels. Many of these were allocated, or placed by the FCC's Engineering Bureau in communities which had previously lacked local FM service. It was also deemed to be a more efficient use of the FM band, finding space for new FM stations in an already crowded spectrum.

As with the AM frequency search process, if your community or area did not receive an FM allocation, you may have to petition the commission to allocate a new channel to your area. This can significantly lengthen the process. It is strongly recommended that if you need to go through the allocation process, you definitely should have a good engineering firm and qualified legal counsel involved throughout the process.

## Noncommercial Educational FM Broadcast Stations

Subpart C of the CFR sets aside the reserved portion of the FM spectrum for noncommercial educational FM broadcast stations. Particular attention should be paid to Subpart C of the CFR 73.501(a) which lists frequencies which are available for noncommercial FM broadcasting:

| Frequency (MHz) | Channel No. |
|---|---|
| 87.9 | [1]200 |
| 88.1 | 201 |
| 88.3 | 202 |
| 88.5 | 203 |
| 88.7 | 204 |
| 88.9 | 05 |
| 89.1 | [2]206 |
| 89.3 | 207 |
| 89.5 | 208 |
| 89.7 | 209 |
| 89.9 | 210 |
| 90.1 | 211 |
| 90.3 | 212 |
| 90.5 | 213 |
| 90.7 | 214 |
| 90.9 | 215 |
| 91.1 | 216 |
| 91.3 | 217 |
| 91.5 | 218 |
| 91.7 | 219 |
| 91.9 | 220 |

[1]The frequency 87.9 MHz., Channel 200, is available only for the use of existing Class D stations required to change frequency. It is available only on a non-interference basis with other commercial educational FM stations. It is not available at all within 402 kilometers (250 miles) of Canada and 320 kilometers (199 miles) of Mexico. The specific standards governing the use are contained in 73.512.

[2]The frequency 89.1 MHz., Channel 206, in the New York City metropolitan area, is reserved for the use of the United Nations, with the equivalent of an antenna height of 150 meters (492 feet) above average terrain and effective radiated power of 20 kW and the Commission will make no assignments which would cause objectionable interference with such use.

## Governance of Operation for Noncommercial Educational FM Radio Stations

CFR 73.503 states: "The operation of, and the service furnished by noncommercial educational FM broadcast stations shall be governed by the following:

(a)   A noncommercial educational FM broadcast station will be licensed only to a nonprofit educational organization and upon showing that the station will be used for the advancement of an educational program.

(1)   In determining the eligibility of publicly supported educational organizations, the accreditation of their respective state departments of education shall be taken into consideration.

(2)   In determining the eligibility of privately controlled educational organizations, the accreditation of state departments of education and/or recognized regional and national educational accrediting organizations shall be taken into consideration.

(b)   Each station may transmit programs directed to specific schools in a system or systems for use in connection with the regular courses as well as routine and administrative material pertaining thereto and may transmit educational, cultural, and entertainment programs to the public.

## CFR Regulating Noncommercial Educational FM Programming

(c)   A noncommercial educational FM broadcast station may broadcast programs produced by, or at the expense of, or furnished by persons other than the licensee, if no other consideration than the furnishing of the program and the costs incidental to its production and broadcast are received by the licensee. The payment of line charges by another station network, or someone other than the licensee of a noncommercial educational FM broadcast station, or general contributions to the operating costs of a station, shall not be considered as being prohibited by this paragraph.

(d) Each station shall furnish a nonprofit and noncommercial broadcast service. Noncommercial educational FM broadcast stations are subject to the provisions of 73.1212 to the extent they are applicable to the broadcast of programs produced by, or at the expense of, or furnished by others. No promotional announcement on behalf of for profit entities shall be broadcast at any time in exchange for the receipt, in whole or in part, of consideration to the licensee, its principals, or employees. However, acknowledgments of contributions can be made. The scheduling of any announcements and acknowledgments may not interrupt regular programming.

### Important Note

Commission interpretation of this rule, including the acceptable form of acknowledgments, may be found in the Second Report and Order in Docket No. 21136 (Commission Policy Concerning the Noncommercial Nature of Educational Broadcast Stations)

86 F.C.C. 2d 141 (1981); the Memorandum Opinion and Order in Docket No. 21136, 90 FCC 2d 895 (1982), and the Memorandum Opinion and Order in Docket 21136, 49 FR 13534, April 5, 1984.

## The FCC Public Notice Process

The FCC requires that if you petition the commission to allocate a new FM channel to a given community you must also state that you will be applying for a construction permit for the facility as proposed. This is when the real "gamble" begins. Even though you invest the money to find an available frequency and you go through the petition process, there is absolutely no guarantee that yours will be the entity selected to receive the construction permit (or CP).

The process is not for anyone who is in a hurry. When this author went through the process, it took 60–90 days to file the petition and have the Commission post it for public notice. Then it sat on someone's desk another 60–90 days after the comment period in case there were any late comments. Then, after another 60–90 days, the commission issued a public notice that it was proposing to allocate the channel, and it took another 60–90 days. Finally, after what seemed an eternity, it was allocated. It was luck that this was the only application. In another proceeding, the author was one of seven applicants and his application was eventually dismissed.

The lesson is that this process can involve considerable time and expense. This is particularly true if the proposal is for construction of a commercial FM radio station in a populated metropolitan area.

You would be advised to propose a noncommercial educational FM radio station, since the demand is less and you are not as likely to attract competing applicants for the frequency. Even better advice is to find a frequency that can later be upgraded to higher power.

Sometimes this can put your application at a disadvantage if someone else applies for higher power on the same channel. However, often you avoid a lot of these political considerations in the realm of noncommercial educational radio.

The Commission still licenses Class D stations: those with power levels of 100 watts or less to cover a school or campus area, but most of them are grandfathered on their present frequency and at their existing power level. A few years ago the Commission voted to discontinue that class, and stations were either grandfathered at that power level, or were required to upgrade. Many stations that were at the 10 watt level increased power to 100 watts to avoid being "phased out" at that classification. Despite this, numerous 10 watt FM noncommercial stations continue to operate across the country.

With the continued crowding and general lack of available frequencies in major metropolitan areas, many people are wishing the FCC would reinstate their acceptance of new applications at the Class D power level (i.e. 10 watts).

It is important to note that these existing "class D" stations grandfathered at a power level of 10 watts must agree to accept any and all interference from Class A stations, according to the FCC's rules and regulations.

However, CFR 73.511(b) states that "no new noncommercial educational station will be authorized with less power than minimum power for commercial Class A facilities. That is, according to CFR 73.211(a)(1): "The minimum ERP for a Class A station is 0.1 kw." This converts to 100 watts. Some people believe that this rule limits the FCC's authority over stations that are operating at less than 100 watts (see Appendix B: Pirate Radio). However, as of this printing the FCC rules under the CFR require that all applications for new FM stations be at least 100 watts ERP (effective radiated power).

## Starting Stations Under CFR Part 15

For the lay person, the rules under Part 15 generally pertain to such things as the amount of acceptable interference produced by a computer or other electronic devices, microwave oven interference, and such devices as remote control transmitters. Other remote control equipment would include remote garage door openers, baby monitors, cordless telephones, remote controlled airplanes, cars and boats, children's walkie talkies, wireless FM microphones, and the like. As you may surmise from this list, the amount of coverage which you can legally reach with a radio station built under CFR Part 15 of the FCC's rules is severely limited in area. The limitation is not measured in feed from the antenna (or radiator), but rather in signal strength.

Under CFR 15.239(a): "Emissions from the intentional radiator shall be confined within a band 200 kHz wide centered on the operating frequency. The 200 kHz. band shall lie wholly within the frequency range of 88–108 mHz."

The essence of this rule is the next paragraph, CFR 15.239(b): "The field strength of any emission within the permitted 200 kHz. band shall not exceed 250 microvolts/meter at 3 meters. The emission limit in this paragraph is based on measurement instrumentation employing an average detector." Said another way, such stations probably will not transmit with much more power than your average remote-controlled garage door opener.

Since FM operates principally as a line-of-sight system, higher antenna systems will yield better coverage. Thus, an antenna mounted on a mast atop a four or five story building will yield better coverage (even at 250 microvolts/meter at 3 meters). If you are considering the installation of an FM system under Part 15 of the rules, you might investigate which building on your campus is nearest the area(s) you wish to cover, and where you might best locate an antenna which will provide the greatest height. Additionally, the antenna must be a permanent part of the device, not replaceable with a larger antenna.

In some areas FM is in a state of confusion and 250 microvolts is useless, but in other areas, with some decent height you can cover hundreds of feet. It should also be noted that FM stations transmitting in mono will provide slightly better range than those using the stereophonic method of transmission.

This is the *only* FM rule for power under CFR Part 15. There are no 100mw FM regulations. It is also important to note that in *all* of these cases you must be using a transmission device which is permitted as being FCC Part 15 Certified, and bears the ID number to prove that. You can purchase any one of several kits and finished transmission devices in the marketplace, but these are illegal under the CFR rules. However, these are few and the commission is working to track down and eliminate them. Every so often the FCC notices these illegal manufacturers and takes steps to prevent the manufacture of illegal broadcast transmitters, according to John Devecka at LPB, Inc. Devecka does believe that the FCC sometimes doesn't act quickly enough in the prevention of this illegal activity.

## Classes of Noncommercial Educational FM Stations and Channels

According to CFR 73.506(a): " Noncommercial educational stations operating on the channels specified in 73.512 are divided into the following classes:

1) A Class D educational station is one operating with no more than 10 watts transmitter power output.

2) A Class D educational (secondary) station is one operating with no more than 10 watts transmitter power output in accordance with the terms of 73.512 or which has elected to follow these requirements before they become applicable under the terms of 73.512.

3) Noncommercial education FM (CE-FM) stations with more than 10 watts transmitter power output are classified as Class A, B1, B, C3, C2, C1, or C depending on the station's effective radiated power and antenna height above average terrain, and on the zone in which the station's transmitter is located, on the same basis as set forth in 73.210 and 73.211 for commercial stations.

4) Any noncommercial educational station except Class D may be assigned to any of the channels listed in 73.501. Class D noncommercial education FM stations applied for or authorized prior to June 1, 1980, may continue to operate on their authorized channels subject to the provisions of 73.512.

There are many other rules that apply to the process of filing an application for a new noncommercial educational (NCE) FM construction permit as well as operating such a station within the guidelines established by the Commission as outlined in the CFR. It should go without saying that if you are not thoroughly familiar with the Code of Federal Regulations (The FCC's "rules") as they apply to licensed and Part 15 low power radio stations, you should seek the assistance of someone who is well acquainted with the rules.

# Chapter 3 ▶ ▶

## FCC Broadcast Applicant Legal Requirements and Other Considerations

### Introduction

This chapter will focus on the legal requirements and considerations associated with filing and pursuing an application for a new commercial or noncommercial over-the-air AM or FM broadcast station construction permit (or CP). Although in rare instances legal assistance from qualified legal counsel familiar with FCC rules and policies might be required in the operation of cable stations, leaky cable, carrier current, and low power (or micro power) broadcast stations (which operate under CFR Part 15 of the FCC rules), these forms of broadcast operation seldom need FCC legal counsel as there is no license required for their operation.

### Part One: Qualifications of Applicant ▶

It should be noted that copies of documents provided by the Federal Communications Commission are included in the Appendices of this book. Specific document titles which might prove helpful in assisting you are:

1) How to apply for a broadcast station
   (FCC Office of Public Affairs, June, 1995)

2) Federal Communications Commission Information about: Low Power Broadcast Radio Stations

3) Understanding the FCC Regulations for Low-Power, Nonlicensed Transmitters

(Office of Engineering and Technology, Federal Communications Commission, OET Bulletin No. 63, October, 1993).

4) FCC Form 301 Application for Construction Permit for Commercial Broadcast Station

5) FCC Form 340 Application for Construction Permit for Noncommercial Educational Broadcast Station

Applicants for new radio station permits must be US citizens and must not have been convicted of a felony. Usually these two major requirements do not pose significant problems. However, should there be any question as to the citizenship of any of the applicants in multiple-person application groups, be sure to consult qualified counsel intimately familiar with how the FCC operates.

## Part Two: Legal Assistance ▶

A note of explanation: throughout this discussion, references will be made to "FCC legal counsel." This is not to be construed that this counsel works at or for the FCC. Rather, this kind of attorney specializes in matters before the FCC. Please keep this in mind when references are made to the term.

### Reasons for Choosing Qualified Counsel

Some applicants have attempted to cut corners and save money by not employing the services of qualified legal counsel who is familiar with FCC rules and policies. This practice is not recommended as it may result in your application being delayed or summarily dismissed for details that might have been discovered after a thorough review by a qualified FCC legal counsel. It has been the author's own experience that the legal branch of the FCC can be very detail-oriented: as noted earlier, simply not crossing your "Ts" or dotting your "Is" can result in getting your application back in the mail requesting that you make the changes, or (worse) having your application dismissed. This can be particularly dangerous if another party has filed a competing application for the same frequency, and their application is complete, accurate and acceptable.

As previously stated, applying for an over-the-air radio frequency can be considered a gamble. If you are in the game to win and you have intentions of getting the frequency for which you are applying, it is best to invest the money and hire a good broadcast law attorney and a good broadcast engineering consultant. Let them handle their respective sections of the application and you will probably encounter fewer problems. You will still have a long wait for your application to be processed, but having qualified professionals on your side is strongly advised.

In some instances FCC legal counsel has been able to help move an application through the process by delivering it in person, getting it hand stamped with a filing date, and in other helpful ways getting things processed a little sooner. Conversely, some applications that were not handled by FCC legal counsel have been misplaced or totally lost.

Take precautions and hire counsel to prepare and file your application. In a situation where you find your application is crossfiled by one or more competing parties, your counsel will already have first-hand knowledge of the particulars about your case and will save you a lot of money in not having to bring another attorney "up to speed" on the particulars of the case.

Another big advantage of having qualified FCC legal counsel is the fact that FCC rules and policies are changing almost daily. Staying on top of these frequently-changing rules is nearly a full-time job and having counsel who knows the ins-and-outs and shortcuts through the bureaucracy can prove to be a major advantage.

## Finding Qualified FCC Legal Counsel

Where can one find qualified counsel familiar with FCC rules and policies? Actually, a good place to start is with the FCC itself. The commission provides a variety of useful documents at no (or minimal) charge to the public. For information contact:

> Office of Public Affairs,
> Public Service Division,
> Federal Communications Commission,
> 1919 M Street, NW, Washington, D.C. 20554
> Phone: (202) 418-0200/Touch Tone: (202) 418-2550

As their June, 1995 document titled "How to Apply for a Broadcast Station" states: "Full details of the licensing procedure and station operation are in Part 1 of the commission's rules, "Practice and Procedure," and Part 73 "Radio Broadcast Services." This includes technical standards for AM, FM, and TV stations, and TV and FM channels (frequency assignments) by states and communities. These rules are summarized in this publication. Copies of the complete rules may be purchased from:

> Superintendent of Documents,
> Government Printing Office,
> Washington, D.C. 20402
> Phone: (202) 512-1800

## Part Three: Preparation of the Legal Section of FCC Form 301 ▶

### Application for Construction Permit for Commercial Broadcast Station

There is a variety of alternative methods to get answers to specific questions. Although word-of-mouth and referral from other station operators is common, one of the best places to check is the trade magazines. Names of attorneys licensed to practice in the District of Columbia who specialize in matters before the FCC, as well as Engineering Consultants whose work is a matter of record with the FCC can be found in such trade publications as *Broadcasting and Cable* and *Radio World.* Referrals for qualified legal counsel can also be obtained from organizations formed to help student media. A list of those organizations can be found in Chapter 14 of this book. One notable example is the National Association of College Broadcasters which has FCC legal counsel available to its members on a limited basis.

The old axiom "you get what you pay for" has pertinence when it is applied to filing for a new AM or FM commercial broadcast station construction permit. One of the greatest advantages of having a qualified professional handling the legal portion of this process is that they work with the rules and documents every day and many of them really know their profession thoroughly. They know which exact wording and terminology is acceptable to the FCC, and they know where and how to use it properly in the preparation of documents and applications.

Some firms do suggest that you, as applicant, can save money by preparing as much of the legal section as you can (or in full), and then have their representative(s) review it for accuracy. In addition, your legal firm will also review the entire application for accuracy. It is very important that you use the exact and acceptable wording on your application. One such example is your statement on proposed programming. Legal counsel familiar with preparing such documents can assist in helping and providing you the correct and acceptable phraseology so you will have an acceptable answer to that section of the application.

Earlier in this chapter there were references to the application process being somewhat likened to a gamble. If you believe this, you will take care in the task of preparing your application for a construction permit for a commercial broadcast station. There is no doubt that the Federal Communications Commission takes it very seriously, as well. Your application will be carefully reviewed under the closest scrutiny by several different departments at the FCC. The two departments of major significance are legal and engineering. This, in part, is the reason that the procedure of processing of your commercial broadcast station construction permit application could take a year or more.

It is particularly important that your application for a commercial broadcast station construction permit be completely accurate in every way. It should be noted that access to the FCC personnel and the various departments has been made more efficient in recent years. The commission's personnel have also reflected an attitude of being more helpful to applicants.

Errors on both sides of the desk are possible, and the commission is usually quick to catch their oversights (or quick to amend/correct their mistakes, if the errors are pointed out in a timely manner to the right person who has the authority to make amendments). Mistakes that you or your legal firm make can be more costly, especially in terms of the amount of time it can take for your application to be processed.

Applications for commercial broadcast station construction permits, particularly if they are for FM broadcast stations in metropolitan areas can be highly competitive. It is an unfortunate fact of life that there is only one permit granted to one applicant.

In the cases where multiple applicants are competing for the same construction permit (referred to as mutually exclusive or MX), any application found to be inaccurate or incomplete will be one of the first to be dismissed. You really do not want this to be your application.

In summary, do not take chances when filing for a commercial broadcast station construction permit. Do your homework and carefully complete your research before selecting the broadcast legal firm and broadcast engineering consulting firm that is right for you. The process will require some time and effort in the form of telephone calls, letters, faxes, or electronic mail postings. However, the time you invest in making your selections will be much less than the time you might lose in preparing and submitting a flawed application to the commission.

## Preparation of the Legal Section of FCC Form 340

### Application for Construction Permit for Noncommercial Educational Broadcast Station

As with its commercial counterpart, the application for a construction permit for a noncommercial educational broadcast station must be complete, accurate, and neat in appearance. All answers and exhibits in the engineering section (as well as throughout the application) should be typewritten and every question must have a statement or some form of answer in every box. This is even true if a particular question does not apply to your type of facility. Leaving any kind of blank on your application without answering the question is an invitation to have your application returned.

Although the noncommercial application form is shorter and somewhat less intimidating than its commercial counterpart, preparing it demands just as much accuracy as the commercial broadcast station construction permit application form. Again, you should notify both the legal counsel and the consulting engineer of your intent to apply for a noncommercial radio station construction permit up front when making your initial contacts. Make sure you have each prospective professional you are considering provide you with a written quotation of what they will be charging per hour or for the project in total before you agree to hire them. The key word is written: make sure they mail or fax you the hourly rate or total cost they will be charging you for this project. You'll probably be surprised that noncommercial applications are often charged at a greatly reduced rate from their commercial counterparts. This is because there is less anticipated risk of having competing applicants for a noncommercial station than there are in the commercial arena. Also, there is often less crowding in the lower, or noncommercial portions of the spectrum. Again, make absolutely sure that, if you decide to go noncommercial, you communicate this fact to these legal or engineering professionals you are considering, and that you get the hourly charges for their services in writing before you begin: both legal and engineering.

As mentioned, some law firms offer the option of having you complete portions of the legal section yourself, saving them time and saving you money. If you elect to complete substantial portions of the application yourself, they may offer a "retainer" agreement where they simply will oversee the project and check it over for mistakes before submitting it on your behalf. Again, however, cutting this corner can sometimes lengthen the amount of time it takes to acquire the construction permit. It should be noted that most of the major broadcast legal consulting firms insist on completing the legal section themselves to avoid conflicts.

A final word of advice in this area: shop around. It is not necessary to employ the best-known legal consultant in Washington unless you anticipate a major battle with a potential competing applicant. Through careful shopping you will find that hourly rates vary immensely. You may even find a firm which will accept your project on a flat fee, although this is somewhat unusual, even with a noncommercial application.

## Working Within Your Circle of Professionals

The questions that will come from your circle of professionals will be simple and straightforward. Be assured that this portion of your application will look professional and will have a minimum number of problems and challenges when it comes under the close scrutiny of the FCC. Typically it is likewise recommended that you have a qualified and experienced FCC engineering consultant as part of your team in this endeavor. Your attorney will likely want to know who is preparing the engineering portion of your application. If you have hired an experienced and qualified engineering firm or consultant to complete the engineering portion of the application, you will find your attorney and your engineer will work well in concert to get you through the process in an expedited manner, at least as much as is humanly possible.

Another reason your legal counsel will inquire about who is doing the engineering is to know whether preparation for possible defenses of the application will be necessary. Typically if a well-known (or at least qualified) engineering consultant whose work is a matter of record before the FCC is being used, there will be a minimum of difficulty. This will also mean savings in terms of legal time to defend against any mistakes in the engineering section.

Again, remember that during the review process, your application will be carefully scrutinized by several major departments at the FCC. This is particularly true of the engineering and legal sections of your application, and it is very important that all exhibits are properly prepared, labeled and included with this document. Any omissions, errors or other oversights can mean a long delay or even possibly result in the dismissal of your application.

It should be reassuring to know that there is a host of people who are prepared to help you find qualified help including this author. In addition to the organizations listed in the appendices of this book, the law firm you select may help you by recommending qualified engineering consultants, particularly those with whom they have worked in the past. If you have first selected an engineering consulting firm, they may recommend legal counsel to help you in preparing your application. The FCC itself, however, will not make any such recommendations as it would be considered a major conflict of interest.

Although it is not required that your legal counsel and your consulting engineer know one another, it can't hurt. Professionals in these circles frequently work together, sometimes know one another, and they all realize that their goal is to help you win in your bid for a new CP.

## Part Four: Filing Your Commercial or Noncommercial Application and Public Notice Requirements ▶

Your application must be complete, absolutely accurate, and neat in appearance. All answers and exhibits should be typewritten and every question must be responded to with some form of answer in every box.

You are required to send the original and at least two photocopies of your application to the following address:

Secretary,
Federal Communications Commission,
1919 M Street, N.W.,
Washington, D.C. 20554

Considering the volume of mail that the Commission's secretary receives, special care should be taken to improve your chances of getting through the crowd of mail. Sending your application and the appropriate number of copies by certified mail is essential.

Applicants for new commercial radio station construction permits are required to publish an announcement in the newspaper of highest circulation in their area. One tip which may save money: the FCC does not require that this be published under the "legal news" section of the classifieds, but merely that it be published in the local paper of highest circulation. You may also be able to publish the announcement under the "public notices" classification of the classified section in the paper, which may save money. Any such announcement must be exactly worded.

When this is published depends on the kind of paper you have in your area. If the paper of highest circulation is a weekly paper, your announcement must be published once each week for the first three weeks after your application for a construction permit is filed. If it is a daily paper, it must appear two consecutive times each week over the two weeks immediately after your application is filed with the FCC. In either case, you must order a "Proof of Publication" Affidavit to prove it was indeed published on the days and dates as specified by the rules. Your FCC legal counsel can provide you with information regarding the FCC rules in this matter, as well as the exact wording for the public announcement. Basically your announcement contents include your name (or name of the group as it appears on the application), what the proposed frequency and power level is, the proposed tower and studio location, and the location where a copy of the application and accompanying exhibits are available for public inspection. A copy of your application and all accompanying exhibits must be made available for public inspection at a well-known public location during business hours. Many applicants are advised to maintain this file in a public location, such as a library or post office. This permits an applicant to make sure the file isn't stolen, and that it is easily accessible to the greatest number of people during normal business hours.

Sometimes interested parties would like to photocopy portions of the application. This is permitted, and charges usually reflect the going per page copy rate available at that location. The public library is one of the best locations for display of your application, as the document can then be placed on noncirculating closed reserve and not leave the building. Often the librarian will also have those parties inspecting the document sign for it, so you can have an idea of the level of interest in the project. However, before doing so, you should consult your legal counsel, as many FCC policies are often in a state of frequent change. It should be noted, however, that nothing can be done with this list of names. It would be advisable to leave the sign-out list in the librarian's hands until the comment period and window for competing applications has closed.

## Part Five: Possible Legal Ramifications from the Crossfiling of Your Application ▶

As previously noted, broadcast station construction permits are valuable and the competition for them can be intense. This is especially true in large metropolitan markets where commercial FM frequencies are still available.

After you file your application, the commission will notify you of its receipt with a postcard. In approximately 60 days you will receive another postcard stating that your application has been tendered for filing. Basically this means that it has passed a preliminary review through the legal and engineering departments and that no major deficiencies or errors have been found.

During this first 60 days it is presumed you have followed the commission's rules in terms of having notice of your application published in your local paper having the highest circulation. When placing your announcement in the paper, it is important that you also request an affidavit of performance from the publisher. Upon completion of the publication of your notice, you will receive the affidavit via first class mail. It is critical that you retain this document for your records, as it is your only proof that the announcement did indeed run on the dates specified, in accordance with the commission's rules. If you do not follow the rules to the letter, a competing applicant who does follow the rules can get your application dismissed on this point.

In fact, any competing applicant can get a copy of your application, which is considered to be a public document, simply by photocopying it at the location where you have placed it for public inspection. The individual who is interested in filing for the same frequency can complete a separate application and file it with the secretary at the commission, as well. In addition, a cross filing party may also take a copy of your application to another consulting engineer or legal counsel. The other applicant's engineer or attorney can review your application in an effort to find inaccuracies, errors, deficiencies or oversights. This is why it cannot be stressed enough to be accurate, legible, professional, and have a qualified broadcast consulting engineer and broadcast attorney review your application to alleviate any mistakes and shortcomings.

## How Will You Know if Your Application has Been Crossfiled?

This is a common question. Typically your Washington, D.C. based legal counsel will be the first to spot any such announcement and will notify you either by telephone, fax, or in writing. Another method (which you can do simultaneously) is to regularly read the trade publications such as *Broadcasting and Cable* which publishes a weekly record of transactions and notices released by the FCC.

You should understand that there is sometimes a delay between the time the publisher receives the information and the time it appears in the publication. Typically this delay period is about a week or two. It is a good idea for you to still monitor the trades even if you have legal counsel on retainer. This is a good double-check, backup procedure just in case your counsel fails to catch a notice.

When you learn that your application has been crossfiled by another party, there are several options available to you. One action is to simply wait for the outcome. Typically the FCC will delay until the window period for comments or competing applications has closed. This can easily be six months or more. If your application is crossfiled, you can expect the entire process to be significantly slowed.

Another option is more complex and will require a significant financial expenditure. In this case, you will have to weigh whether the investment is worth the outcome. You can follow the same procedure outlined earlier: that is, get a copy (from your legal counsel or from the other applicant's public inspection file) and have your attorney and consulting engineer review the application(s) for deficiencies.

At that point you can request that your legal counsel prepare and transmit a letter which seeks to have the competing application(s) dismissed. At that point the decision is in the hands of representatives of the FCC.

If neither or none of the applications is found to be defective, the pending applications are then handed over to an Administrative Law Judge for comparative hearing. Each applicant is required to deposit a fee of $6,000.00 prior to the hearing. Legal counsel is not mandatory, but would be strongly encouraged if there are mutual applications for the frequency which you or your group wants.

There are always other options. One such alternative is to find another frequency, if possible, in the same market or area. Sometimes a clever and creative engineer can facilitate this by proposing that other stations on adjacent channels change frequencies, thus making additional space available on the spectrum so that both applicants can be granted a construction permit.

Another option that exists is negotiation. There have been cases where one applicant agrees to "sell out" and withdraw an application. This is sometimes done in exchange for something of value, such as financial consideration or perhaps an existing station or construction permit. An important point to note is that the FCC does not take "trading" in radio and television station licenses lightly. The basic premise is that a licensee is granted a license or permit to construct and operate a broadcast station to serve the public in the public interest. In the case of a station sale, typically everything is assigned a price, such as equipment, facilities and even community goodwill; everything but the license. The rules do not permit the assigning of a monetary value to the license. And license transfers must be approved by the Federal Communications Commission.

Typically a license transfer takes about 60 to 90 days barring any unforeseen circumstances. Exceptions to this are when the commission receives objections to a transfer, when there are mitigating circumstances, or when application is made for a transfer in the case of stations that are deemed to be in distress (usually financially). In the latter cases, the commission can make a determination in as little as 30 days.

The options of negotiation or finding an engineering solution are typically preferred to paying the fee and litigating the matter before an FCC representative. However, when there are many applicants for a single frequency and the stakes are high or there are no engineering options, a determination from an Administrative Law Judge (ALJ) may be the only alternative. However, the FCC is not presently holding competitive hearings.

## Part Six: New Legal and Rulemaking Considerations Associated with Noncommercial Educational Broadcast Applications ▶

The major differences in preparing and completing the engineering sections of commercial vs. noncommercial broadcast station applica-

tions were discussed in chapter two. Much of the information requested in the legal sections is similar, as well. For your convenience, copies
of each type of application are included in the appendices of this book.

# Chapter 4 ▶ ▶

## THE PLANT:
### FACILITY REQUIREMENTS/
### PHYSICAL SPACE:
UTILITY/VENTILATION/
SECURITY/SOUND
CONSIDERATIONS

---

## Introduction

There are several terms used when broadcast investors, brokers and management types gather. The term used to refer to all of the physical elements needed to produce your product is the "plant."

In the newspaper business, the plant would most certainly include the presses. This would also include all of the hardware and assets necessary to conduct business on a daily basis, including office furniture, desks, filing cabinets, computers, telephones, fax machines, copiers, and all of the other fixtures used. The plant may also include the building, if owned or leased, as well as the land on which the building is situated, if it is an asset of the company.

In much the same way, radio stations are considered to have a plant. A station's plant also includes these same items outlined above (excluding the presses), but would also involve all of the electronic devices used to produce programming and to transmit a quality signal over a specified geographic area.

## Testing the Waters/Preparing for Market Entry

One of the greatest advantages to starting a noncommercial educational station is that you are making the choice to not compete with the commercial counterparts in your area. However, you should be aware that some local commercial station operators in your area may still not welcome your station in what they may consider to be their market. The mere fact that you are going to be attracting an audience which, may in turn, affect or diminish their numbers can create ill feelings toward your efforts. Proceed carefully, practice using your best inter-personal communication skills and be the epitome of tact.

On the positive side, you will not be competing directly with them for dollars. However, depending on how aggressive your stance is on generating dollars from underwriting, you may still have some effect on the local radio revenue generating economy. It is recommended that you approach station managers in your area with a positive outlook and strive to build good relations with them from the beginning. If they learn early on that your project should not be perceived as a threat to their existence, you will be leagues ahead in the long run. They should be convinced of your intentions, and that, if nothing else, your station will be a possible source for future announcers.

You should be cautioned that it is illegal to engage in such prac-tices as "rate fixing" in your market. Rate fixing is the practice of agree-ing with other station managers in your market on one set of rates for all stations in the market, which is in violation of fair trade agreements and other rules.

If you were considering construction of a commercial station, you should steer clear of such endeavors. However, co-promotion of events, particularly those on your campus can offer a delightful opportunity for already established stations, and they can be helpful in co-promot-ing the launch of your new station, as well. In addition, sometimes these stations can help your station by donating equipment.

## Pre-construction Planning Considerations

In some respects you are adding to the communications family of your school or organization. Like a family, you and your group must make careful and thorough plans to accommodate this new addition. You should try to anticipate every possible problem and ways to answer the challenges.

## Security

Broadcast equipment can be very expensive, particularly the new state-of-the-art non-linear digital editing equipment technologies. It is important that you plan ahead and consider how you can restrict access by only those persons on your staff who have authorized entry into your studios.

Of particular concern is after business hours and weekend access, where your regular clerical and/or administrative staff are often not on site. The temptation can be all-too-great for an announcer to invite one or more unauthorized friends or guests up to the station. Even with strong policy handbook guidelines, these kinds of rules can be broken and are difficult to enforce. At minimum you should consider restrict-ing access from outsiders, and giving evening, overnight and weekend announcers the opportunity to determine the identity of those wishing access into the station during non-business hours before the door is opened to them.

Many campuses have security on patrol at night and on weekends. If a check on the station's locks can be added to their routine patrol, this can help. It is also important that your evening, overnight and weekend announcers all have the knowledge of how to contact campus security or 911 personnel in the event of emergency situations.

Passing out keys or the combination to your station's exterior locks can always pose to be a problem. Some university campuses have "DO NOT DUPLICATE" stamped on the keys, which can deter those students who may try to have duplicate keys made without proper authorization. Even so, sometimes keys and combinations are passed around to unauthorized people. The best protection is to acquaint your staff with the fact that security is everyone's business. It is for the safety of your staff and station that these rules be strictly enforced.

Some stations find it necessary to chain and lock down important pieces of equipment like computers and consoles. Even bolting down consoles, cartridge machines and the like can slow down and sometimes prevent burglary.

Your station should be located in a place where you can easily control access. It is preferable to locate it behind two sets of locked doors, such as a room within a larger building, so it becomes necessary to have two keys to gain access into the station. Although somewhat cumbersome, it can likewise improve security by having locks that are keyed differently, such as doors to administrative offices different from those accessible to announcers, production personnel, engineers, and other staff members. It can get complicated, but a series of submaster and master keys can help those with multiple levels of access gain entry into those rooms for which they are authorized.

Regular annual or semi-annual changing of the locks can also improve security and deter unauthorized access to station facilities by outsiders, as well as by former station personnel. Of course, having a night watchman checking pictured identification badges is preferred, but is a luxury not available at many student stations. A check with your campus security office can help you. But prior planning of where to locate your studios with security in mind can also prove to be advantageous.

## Acoustic Insulation

In addition to security elements, another consideration for your studios is isolation from external noises. Common sounds include planes, trains, busses, trucks, cars, trolleys and the like. Other sounds can be bells that may ring between classes, PA system announcements, students talking and slamming lockers in the hall, people knocking on the station's studio door(s), telephones ringing, and even blowers for ventilation ducts.

When you think you have found the perfect room for your studio(s), sit down in it and see if any external noises can be detected. If so, it might be worthwhile to either pick a different room location or, if this is not an option, think of ways to isolate this room from the outside world. Keep in mind, though, that acoustic insulation can be expensive, especially the adding addition of walls and doors to act as sound locks.

Your school may not be able to budget such extravagances, and you may be unable to convince your school's administrators of the importance of keeping noise out of your studios. Under these circumstances, you simply must do the best you can with what you have. However, it's much easier to relocate a studio that is in the planning stage than after it is already built, wired, and on the air.

## Heating and Cooling

Maintaining a comfortable work environment is essential to the success of any business that involves people working within a structure. Depending on your geographical location, this can involve a great many variables.

The summer sun can make life unbearable if your office is located where the blistering summer sun can shine directly through your windows or if your walls have insufficient insulation in southern states. Conversely, you can be very uncomfortable in the winter in northern states if you are working near a wall or window which does not provide sufficient thermal protection.

If you have a choice, select a building which provides adequate protection from the elements. This should go beyond such obvious considerations as a leaky roof or poor structural integrity. An example of this scenario was the offices of a church in one of the southern states. Some of the offices for the church were located along a southern exterior wall. In the summer on hot, sunny days the temperature would soar in these offices because of large ceiling-to-floor windows. The pastor and some of the other workers would adjust the thermostat so the air conditioner would keep the temperature in these offices at a comfortable level. However, the workers in interior offices were always cold because their offices were not compensated by the heating effect of the sun through windows.

Some of these considerations may not be obvious until you build your station and begin operations within the walls. In any event, if you are building a new building and designing the layout of studios and offices, the location of the thermostat is paramount to maintaining comfortable environmental working conditions. Typically locating the thermostat on an interior wall in a hallway that is accessible is preferred. Before you build it is advisable that you consult a certified, professional heating and cooling contractor.

## Ventilation

Another important consideration in determining your studio location within an existing building is one of quiet ventilation. It is important that you check to make sure that each room in your station has proper and ample air flow. This usually means ductwork such as a register for forced air into the room, as well as a return air flow duct back to the furnace or air conditioning system. If you do not have return air flow, studio doors can suddenly begin closing or opening by themselves as the air attempts to rush past them.

Proper quiet ventilation for each studio is also an important consideration for the comfort of your personnel. After working four or more hours in a stuffy room, anyone can become edgy and irritable. Plan ahead and make sure your rooms are adequately supplied with sufficient, unrestricted air flow that remains quiet when it is circulated.

## Lighting

A properly lit studio is important. Track lighting to create a mood, unless you have the budget and time to install it at your school, can be a luxury. Adequate lighting, however, is important to provide for your staff. Studies have suggested that natural light can be the best, but sufficient lighting of an artificial variety is acceptable. Still, windows can prove to be an asset if your announcers have the opportunity to see outdoors while they are on the air. They can provide listeners with immediate information as to what the weather is doing, for example. This saves them the trouble of having to run down the hall and look out the back door to see whether it has stopped raining or not.

Track lighting can accentuate your recording console and various elements of your studio. However, it is seldom found in student settings. What is more common and prevalent is standard fluorescent lighting.

Whether fluorescent or incandescent, lighting can help to improve or detract from a mood. There are numerous energy-saving lighting options available on the market.

For example, regular incandescent bulbs can be replaced with Earth lights or twin-tube compact fluorescent lamps. These fluorescent replacements have a screw-type base, produce a warm white light but use less energy. Earth lights have fluorescent tubes that are shielded by a plastic diffuser for a light more resembling incandescent. They use about 76 percent fewer watts (18 versus 75). Twin-tube compact fluorescent lamps also produce a pleasant, unobtrusive light and use less energy (11 watts instead of 40 and 13 watts instead of 75). Some people also believe that these kinds of lamps seem to outlast their incandescent counterparts. Another obvious advantage is that you can easily replace your old, existing incandescent lights.

As a general rule, experts recommend that your light source should be located approximately 30" above your work area. Whether constructing a new building or remodeling to accommodate your studios, it is strongly advisable to consult a lighting or decorating expert to determine the best combination of lighting for optimum efficiency and staff morale in your station.

## Utility Space

When you have considered all of the obvious requirements for space, it is then important to think about your needs that are not quite as obvious. In addition to the obvious spatial requirements for studios and offices, you must also allow utility rooms and closets. In much the same way that a house needs cupboards and closets, a radio station needs storage rooms and utility rooms. These rooms must be considered for the optimum location of such things as your heating and cooling equipment for the most efficient operation.

In addition, you must design your radio station to include utility closets for such things as people's coats, equipment storage, cleaning and maintenance supplies, and so forth. These utility spaces are very important considerations and must not be overlooked in your overall plan.

It is also wise to plan on a room to hold staff meetings and conferences, and perhaps client presentations. The size of the room is traditionally based on the size of your staff and your own needs for the space.

In the author's own experience in building design, it was fortuitous that he considered additional space for an engineering and equipment maintenance and storage. This room could be locked for security purposes and was invaluable as a space where the station's engineer could work on equipment undisturbed.

## Summary

It is important that you take time to research and consider the many alternatives that go into your station plant. You will be operating in it for many years. And many times the morale and positive achievements of your station's staff can be reinforced with a good plant and optimistic, upbeat and cheerful surroundings. Visit other stations, talk with other people, request manufacturer's literature, and explore all of the options within your budget. In the next chapter we will explore what goes into broadcast studio design, and what may work best for your operation.

## Conclusion

Do not spend beyond your budget and be sure to work within the parameters established by your school's administration and physical situation. Plan your station with an eye toward the future and a multiplicity of possibilities.

There are always "little" things that you can do to change the working environment in an existing building. These include changing the surroundings, installing window blinds or curtains, or even changing your lamps from incandescent to fluorescent simply by changing bulb types. Relamping some of your fixtures can result in savings and improve the appearance of the station environment.

If you are on a tight budget, take advantage of local talent in the form of volunteer contractors and carpenters. You also have a large pool of volunteer help, so plan to use it at every available opportunity.

# Chapter 5 ▶ ▶

## STUDIO PLANNING AND DESIGN

### Considerations and Planning

Careful consideration must go into station planning and physical layout of your studios. Since a great many people will be spending a lot of time in them, it is worthwhile to take extra time to plan the physical layout of your station's studios. In planning, a lot depends on what you are going to be doing, and whether you intend to expand your operation or not.

A studio and entire radio station can fit on a desktop computer. However, physical comfort of an operator who might have to sit in the same room should also be considered. A closet might hold two turntables, a console, telephone and a cart machine (as well as one person), but working in such cramped quarters can be hard on a person's morale and physical comfort.

Another consideration is whether your station will be commercial or noncommercial, and whether you plan to have a lot of commercials or underwriting announcements produced. If this is the case, you might consider having a live on-air studio and a second studio to be used for production purposes. If you plan to program a lot of local news or sports, you might have a news room and news booth, as well. Your plans will depend on what you are trying to accomplish with your programming overall. It is advisable to try to allow for and accommodate the unexpected, and to plan for more than enough room that you currently need. It becomes very difficult and costly to add space after you've already constructed the walls.

### Physical Comfort

Many professionals believe the human voice is best projected from a standing position. You might consider a stand-up console in this event. However, many schools must also adhere to rules regarding accessibility. In the case of a physically challenged person in a wheelchair, a console which requires the announcer to stand could pose a problem.

Along the same line of thought, you might also consider making the width of your studio doors sufficient to permit wheelchair access. You should also check to assure accessibility into your building, studio, and control room for the physically challenged.

Ventilation is a very important consideration. Keeping the air quietly circulating can sometimes prove to be difficult. Placing the thermostat in a location where the temperature in your main studio or control room can be controlled comfortably is also an important consideration. Heating as well as air conditioning is important as the weather and the seasons change. It is essential that you have plenty of airflow, particularly since much of today's solid state equipment requires sufficient air cooling and space around each component in the equipment rack. Take care not to locate equipment racks, cabinets and studio furniture in the airflow path. Large windows and small rooms can cause drafts, block airflow, and make rooms unseasonably uncomfortable. Recently a station manager reported that it was difficult to keep his building comfortable for employees. If the physical facilities are already built you are going to inherit these kinds of challenges.

## Security

Accessibility is important. Controlled accessibility is even more important. You want your administration and staff to have easy access to the studios, but you also want a way in which the studio complex can be secured after hours and on weekends, or over holiday periods when the station(s) may be off the air.

## Soundproofing/Acoustic Insulation

Make every effort to locate your studio(s) in a location free of exterior noises. This is particularly important where busy hallways full of students may generate a lot of noise during class transitions.

It is important that your station maximize visibility, but not to the point where distracting sounds can be heard over the air. In addition to properly damping the ventilation system, it is important that you install sufficient acoustical insulation to effectively keep outside sounds from entering the studio, and studio sounds from being heard outside the room.

Recently the Operations Manager of the local public radio station told one of this author's classes that there is no such thing as an acoustically-insulated studio. He may be right. There are several excellent books on the subject of building an acoustically-insulated room. These include offset studs, floating floors and ceilings, avoidance of corners and sharp angles, and tilting glass windows down to reflect sound toward the floor.

Many student radio stations have carefully insulated walls. There are several different kinds of acoustic insulation on the market. One of the most commonly used brands is called Sonex. The product has several competitors offering lower prices, but Sonex still seems to be the acoustic insulation of choice for many schools, NCEs (noncommercial educational FMs), and commercial television and radio station operations.

## Studio Size: Is Bigger Better?

How big should my studio be? Should the production studio or recording studio be larger than master control, or the control room? These are questions that are often asked, but difficult to answer. It would be too easy to say that the perfect small NCE radio station control room should be at least eight by ten feet. In reality, no one can actually determine how large your studio should be until you can describe what you plan to put in it.

Essential questions to be answered include: "What will this studio be used for in your planning? What kinds of equipment do you plan to install? Will you ever use the room for multiple-person programs?"

## Planning and Designing the Small NCE Station's Control Room

Determine how you plan to use your room and what/who you plan to put in it before you decide on its size. Although there are no "rights" or "wrongs" in planning, there are some general guidelines that you can follow to make things easier. For the purpose of this discussion, we will assume you are planning to build a small studio and your project is on a very tight, limited budget.

Time is money. The time that you invest in planning your design prior to construction will be well worth it. Get as many opinions as you can. Do not rule out getting suggestions from the people who will be working in the facility, even if it's only for a semester or two.

### Twelve Suggestions for Small Station Control Room Studio Design

#1:   Get as much input from as many people as you can.
#2:   Determine how many people you'll ever have in the room.
#3:   Physically lay out all your equipment and get a feel for how much reach space you will need.
#4:   Have everything within the announcer's reach.
#5:   Locate what you will be using the most nearest the console.
#6:   Consider appearance of your studio furniture.
#7:   Consider alternatives to expensive acoustic insulation.
#8:   Make equipment easily accessible for maintenance.

#9:     Use ON-AIR lights.

#10:    Install a quiet door, insulate the back of it and have a window in the door or next to it.

#11:    Try to have the console positioned so the announcer can look toward the door.

#12:    Position the control room so the announcer can actually see outside, if this is at all possible.

Suggestion #1: Get as much input from as many people in your school as you can. It is surprising how many creative ideas will spring from a group of interested people.

Suggestion #2: Determine how many people, at maximum, you might ever have in the room. Typically small station control rooms are designed to be used in a "combo" operation. That is, the DJ or announcer also operates the control board. With this in mind, locating all necessary equipment within easy reach is important.

We mentioned earlier the issue of whether to build a stand-up or sit-down console. The distance across the console and the number of people who may be working in front of it at any given moment will often determine the lateral width where equipment will be located on each side.

Suggestion #3: Physically lay out all your equipment and get a feel for how much space you will need. Placement of your equipment might also be affected by how it is made. Some broadcast gear is made to be rack-mounted. Standard broadcast equipment racks accommodate gear that is exactly nineteen inches in width. Some equipment racks are seven or eight feet tall, while there are also three or four-foot high units. Placement of your equipment rack, should you decide to use one, should not block the view into another studio, an exterior window, or the transmitter room.

Other equipment is made to sit on a desk top. Some turntables sit on a counter top, while others (such as old Gates and Russco Studio-Pro models required that holes be cut in the counter top.

Suggestion #4: Have everything within the announcer's reach. Start by determining reach distances required for an announcer to push buttons on the equipment. Inventory the sizes and controls located on all of the equipment planned for the room. Determine if it will require rack-mounting or will sit on the counter. Measure, physically place the equipment around you, and determine the easiest and most attractive way to arrange it, while allowing for comfort and accessibility for the combo announcer. Sometimes buttons can be wired to start reel-to-reel machines that are beyond the reach of the announcer.

Suggestion #5: Locate what you will be using the most nearest the console. If you plan to have your underwriting announcements and other messages on cart, locate your cart machine immediately to the right or left of the console. If you plan to use CD players more often than records, place the units near the console as well. Typically, if two

CD players are used, they are located side-by-side. Try to avoid having them on a shelf where the announcer would have to stretch to reach a compact disk player and run the risk of dropping disks.

Suggestion #6: Determine how your studio furniture will appear. A carpeted counter can help to keep sound down, minimize dust floating around the room, and may prevent a lot of records and CDs from being scratched. It will also hold up well under a lot of abuse, where Formica or veneer-covered counters might be scratched, written on, or cracked. Since smoking is not allowed in most studios, there should be no problem with anyone's cigarette burning a hole in the carpet.

Suggestion #7: If you are on a tight budget and are not able to afford some of the more expensive acoustic insulation products that are on the market, there are alternatives. Some stations have successfully used cork tiles to cover the walls, while others have used short nap carpet. Even egg cartons painted flat black and stapled to the walls can be effective, although this might look somewhat less professional. It is important that you cover the walls with something that will acoustically insulate sounds from entering or leaving the room, but this can be tastefully done in a wide number of ways.

If your budget permits the purchase of some Sonex or another similar acoustic insulation product, you can still minimize expenditures by covering only the top half of the wall. Often your counter tops will insulate the lower part of the room. Tape cartridge racks can be mounted on the walls, so the flat space is minimized further. Large studio windows with double Plexiglas and a trapped, dead-air space in between the thicknesses are effective. Safety glass will serve much better, but the Plexiglas is cost effective and, as long as the dead-air space between the sheets remains sealed and no one scratches the Plexiglas, it can be sufficient.

Each thickness of glass should tilt out at the top and in at the bottom, so any sound waves reaching it are reflected toward the floor. Wood paneling could be used on the back walls, offering an improvement over smooth, reflective surfaces. A suspended ceiling consisting of acoustic tiles can further help to deaden the sound.

Suggestion #8: Be kind to your engineer. Plan to install your console and other equipment in a manner that is easily accessible for maintenance, repair and emergency replacement.

This author once designed a control room to permit the announcer to look out over the city through large ceiling-to-floor windows from our second-floor offices. Since the room was narrow, the countertop was mounted right up against the windows. This made an excellent observation point for the announcer and allowed passers-by to see the announcer in action. We also had a sign which would light up whenever the announcer went on air, and the sign was large enough to be easily seen from the street below the studio windows.

Wiring for our console had to be accessed from the back of the board. If you have ever stood on your head and soldered for six straight hours, then you can certainly understand the intolerable pain which the two installation engineers (the author was one of them) had to suffer.

Learn a valuable lesson from the physical discomfort that these engineers endured: if you have the room, allow three or more feet behind all of your equipment for easy access. This may not be possible in all of your studios, but it certainly should be in your main control room and perhaps your largest (if you have more than one) production studio. Both you and your engineer will appreciate it for many years.

Suggestion #9: Use ON-AIR lights. These warning lights will alert others and keep them from barging into the studio when the announcer is on the air. Make it a station policy to keep the door closed when the ON-AIR lights are illuminated.

Suggestion #10: Install a quiet door, insulate the back of it and have a window in the door or next to it. Whether you carpet the door or use Sonex or another insulation material, keeping sound from coming through such a large, flat surface can be a challenge. Some stations use a double-door arrangement which is known as a sound lock. Having to open and close two doors can prove to be inconvenient for staff members, but it will definitely reduce the noise.

It is also useful to have a window either in the door or next to it so that people can see if the announcer is in the room, whether someone else is in the room with the announcer, and whether it is "safe" to enter. You can always hang a curtain over the window if the announcer requires privacy or does not want to be observed. However, the window has saved a lot of "collisions" from occurring as staff members hurriedly come into and leave the studio during hectic times.

Suggestion #11: Try to have the console positioned so the announcer can look toward the door. The announcer should not be surprised or startled when someone enters the room. This is especially true when there are a lot of visitors or strangers in a station.

Suggestion #12: Position the control room so the announcer can actually see outside, if this is at all possible. There is nothing worse than hearing and describing sunny skies. when it is pouring rain outside. This may not always be possible, particularly if the station is assigned to certain rooms within a building. However, if possible, this need should be accommodated. The ability to see outside and have natural light can also have a positive psychological effect.

Working within already established structures can be frustrating and very limiting to a studio designer's creativity. However, you may be forced to work within the boundaries that are already established. First try to be appreciative of whatever space is made available for your new station.

Recently a young student radio station staff member posted a message via e-mail to the author. After several years his station was being moved a location next to the school's steam room to one near the lobby of the student center. Sometimes it takes time for your new station to build credibility. This credibility-building process can sometimes be enhanced by programming such as a "President's" or "Principal's" corner, where the leader of your school can come into your studios and tape or present a weekly show.

Perhaps if an administrator has to go into a room next to the school's steam room to do his or her show, changes might soon be suggested. Good things can happen if you apply grease to the right wheels. Always maintain and project a positive image within your community and outward to your audience.

## Planning and Designing the Small NCE Station's Production/Recording Studio

This action is very similar to the suggestions made for designing the small NCE radio station's control room. From a practical standpoint, the steps are much the same:

### Production/Recording Studio Design

| | |
|---|---|
| #1: | Get as much input from as many people as you can. |
| #2: | Determine how many people you'll ever have in the room. |
| #3: | Physically lay out all your equipment and get a feel for how your studio will look. |
| #4: | Have everything within the announcer's reach. |
| #5: | Locate what you will be using the most nearest the console. |
| #6: | Consider appearance of your studio furniture. |
| #7: | Consider alternatives to expensive acoustic insulation. |
| #8: | Make equipment easily accessible for maintenance. |
| #9: | Use RECORDING lights. |
| #10: | Install a quiet door and insulate the back of it. |
| #11: | Try to have the production console positioned so the announcer can look toward the door. |
| #12: | Position the production studio for privacy. |

Generally speaking, many of these suggestions for construction of a production studio are similar to those for building a control room. The production studio should be slightly larger. This is to accommodate the additional equipment you may already have and which you may obtain in the future.

It should be noted that all consoles are different. On air consoles are typically different from their multi-track analog counterparts. Multi-track consoles and analog open-reel recorders also may require additional physical space. This is particularly true if the reel-to-reel machines are mounted on roll-around carts.

Some NCE/student stations are now moving into the digital age. Non-linear digital audio editing on multiple tracks is rapidly replacing analog equipment in student stations which receive good economic support.

Digital consoles are typically represented graphically on a computer monitor. With CPU sizes being reduced and hard drives being made physically smaller, yet with higher capacities, the physical space requirements for a digital audio studio can be appreciably less than for a multi-track recording console. This is particularly true with some of the older mixing consoles, some of which had as many as sixteen input modules.

You will need to design your own production studio in a manner that is both appealing to the eye and effective from an operational perspective for your particular equipment.

If your station is to ever have any kind of interview or talk program, such as the "President's Corner" idea suggested earlier, or perhaps to accommodate live bands for recording purposes, you might also consider building an adjoining second studio with table and chairs and multiple microphone jacks in the wall between the studios.

## Announce Booths, Auxiliary Studios

If your school or organization has the available space, you might consider additional studios for recording or interview purposes. If your station is a "full service" station where live newscasts and sportscasts will be presented, you might also consider booths for these. For maximum convenience and flexibility, the news and sports booths should adjoin the news room or wherever the news and sports broadcasts are prepared.

This makes last minute leaps for forgotten carts a lot shorter, rather than having your announcer running down the hall and knocking people over in a mad dash to get what is needed before the cart that is on the air ends.

As you can see, a lot can go in to planning, designing and building new studios. It is important that you be open to suggestions and have everyone who is involved in the design process try to envision both present and future needs. It is also important that they envision what catastrophes could occur with inappropriate designs.

Experienced broadcasters who have worked in a variety of stations can be very beneficial to your cause. Seek their advice, draw out the design, tape off the floor and put the equipment within the marked-off area. Then, envision yourself and your announcers working within that area.

Thought, time, creativity, and a multitude of opinions will help you achieve an effective design for your NCE/student radio station's new studios. If you really have connections, see if your school's CAD class can help you with the designs and blueprints.

## Conclusion

To the earlier question "Is Bigger Better?" in terms of NCE/student radio station studio design, the answer is: slightly bigger is better. Too big can be too much, and you can lose that magical feeling of intimacy that only radio can provide. If your studios become real "showplaces," you might even consider entering a contest such as the NACB "Station of the Year" competition. Good luck, and please do not hesitate to contact this author for help and input.

# Chapter 6 ▶ ▶

## COMMUNITY SUPPORT AND FINDING RESOURCE PERSONNEL

In some respects starting a new NCE (noncommercial educational) radio station is a greater challenge than starting a commercial radio station. One of the greatest challenges in successfully attaining your goal is to not simply get a radio station on the air, but to plan on keeping it going successfully for many years. When a license is granted, you assume an awesome responsibility. In the words of the Communications Act of 1934, you have agreed to serve your community with programming that is in the "public interest, convenience and necessity." You must give something to your community if you ever expect to receive support in return.

Community support can come in a variety of ways. These include human resources, a loyal core group of listeners who can be motivated to action by your programming, and financial resources. Without giving your community something, you should never expect to get anything in return and you might just as well stick to amateur radio or the citizens band.

## Programming

Providing variety programming that people want to hear can be a very demanding task. Programming a radio station requires talented, creative people who have a clear understanding of their mission and the people that they are attempting to reach with their station's programming.

Many public radio stations provide listeners with a monthly program guide which helps their audience know when to tune in to hear specialty programming, special music features, and other unique programming offerings. This service helps to build a station's core audience and provides listeners with a sense of security in knowing the station plans ahead and has the best interests of the audience at heart.

## Locating Qualified Personnel

Many public and community radio stations depend on volunteers and sometimes interns to staff the station. People with a strong loyalty for the station (attentive listeners) are the commodity that a station can provide to a potential program sponsor. Ironically it is people that can prove to be a station's greatest asset. In the decades of the 60s and 70s personality radio was at its peak. Many listeners tuned to a particular station to hear their favorite disk jockey. Many of these personalities had their own identifying trademark on the air. One such example was Wolfman Jack's gravely voice and sense of humor. Other personalities had horns, bells and whistles or talked fast. Many of these radio personalities ran a very tight format with absolutely no pauses or dead air.

The days of strong personality radio have largely passed into history. There are still noteworthy individuals like Larry King, Rush Limbaugh, Howard Stern and Paul Harvey who attract large audiences. Although some national personalities still exist, their numbers have diminished through the years. In these times listeners are likely to tune to a particular station for its programming more than they will for a local personality. There are still exceptions in some markets, of course, but programming is now the main attraction for audiences.

Your station's programming is very likely to come from people. People will most likely do the audience research to get an idea of what listeners want to hear. People will be answering the telephones, talking on the air, gathering your news, sports, weather and community announcement information, and even filing the records and calling on area businesses as possible sponsors.

Unless you plan to have your station totally automated or completely satellite-driven, it will take people to run it. People, especially talented people, will be costly for a commercial station. Typical payrolls for small to medium market stations can easily run in excess of six figures annually. When you add workman's compensation insurance, benefits, and payroll taxes, it is necessary that you to start thinking about how your new station's budget is going to be able to support the staff that will be required to operate it.

That is why many noncommercial educational and community stations rely on volunteers to help them. On college campuses there is available and willing talent eager to go on the air or to help your station in a multitude of capacities. Even in small communities, there are always people who are attracted to the medium of radio who are willing to volunteer some of their time in exchange for the opportunity to learn more about radio.

Some departments in colleges and universities also offer college credit for students who intern at a station. These arrangements can sometimes be negotiated with the school faculty or administrators. There is also the possibility of "articulation agreements" where area schools are willing to offer credit to individuals who volunteer to intern at your station. In nearly any community there is always a pool of talent; the challenge is how to reach these people and get them actively involved in the operation of your station.

Recruiting new people can be a challenge for a new station, particularly if it is not yet on the air. In these cases, you may have to rely on your school's student newspaper, fliers, word of mouth, student organizations, in announcements, and sometimes even announcements on local cable channels. Look to your staff for new and creative ideas to both promote your new station and to recruit more help.

Sometimes retention of qualified volunteers and staffers can prove to be an even greater challenge. You must be cautious of the possibility of "burning out" those people who step forward first. Similarly, you must also continue to challenge volunteers in new ways, or they will quickly become bored and not see the value of spending their time to help your station. The keys to successful volunteer staff motivation are to activate and motivate. You must be creative and ingenious in your efforts to keep everyone excited about what you are doing and very aware that it is their effort that is keeping the station vibrant and effective in reaching its goals.

This readily available pool of volunteers that is available to college, high school and community NCE stations can be vast and very talented. Locating these people and keeping them challenged in new and creative ways while tapping their creative energy can be a very exciting task. In any event, do not waste the creative potential of those who volunteer or intern at your station. Many student stations are very open-minded in their operation, and the doors are often left open to new promotional and programming ideas that will please their audiences. Then again, sometimes these people want to be paid for their efforts.

## Financial Resources

Finding sufficient money to keep your station on the air can be a major problem, especially depending on how small and just how new your station is. If your station is affiliated and funded by a major school, you have a definite advantage. However, if you are starting a new community station or if your school has a very small or non-existent budget, you should seriously consider alternative methods of funding.

Many public stations openly solicit their listeners for ongoing financial contributions to support their operations. In addition, two or more times per year these stations hold fund-raisers. These on-air events require additional volunteers to take calls and often involve heavy promotional consideration. Sometimes donations are sought from area businesses, then the station auctions off the donated items.

Some managers have referred to these types of "telefund events" as "beg-a-thons." However, such fundraising events can be very successful. This is particularly true if the calls are carefully recorded in writing and there is a method in effect to track the progress of donations as well as to make follow-up calls to donors.

Another method of soliciting financial support is through the sale of underwriting announcements and program sponsorships. An advantage of noncommercial stations is that they can "sell" underwriting messages and program sponsorship announcements in much the same way that commercial stations sell spot announcements to business clients. The sad thing is that some noncommercial stations steer away from this practice because of their fear of breaking the rules when it comes to the copy for underwriting announcements.

The simple "DOs" and "DON'Ts" of the applicable underwriting rules are covered in Chapter 12 "Underwriting and Sales" in this book. Questions that are specific to underwriting and sales are further discussed in Chapter 15. Great successes have been achieved in selling underwriting announcements and program sponsorships if the station copy writers are careful and know the rules regarding what is legally permitted and what is not.

Many NCE and community radio station operators do not realize that they can air "full-blown," unrestricted commercials when they are advertising a non-profit business. However, it is important that you are absolutely certain that the business is legally registered and not-for-profit. In addition, individuals can make pledges or sponsor specific programs, which can prove to be another noteworthy and lucrative income stream.

## Public Radio's Financial Support

Government support for public radio was established with the Public Broadcasting Act in 1968. From that time until the present, a major portion of the financial support for public broadcasting and public radio stations, in particular, came from federal sources. Specifically the Corporation for Public Broadcasting made grants available to full-time, full service stations which had staff sizes totaling 14 or more paid employees.

Today, however, this picture is changing. Federal government funding is drying up. Many public radio stations are realizing they must rely more on their own fundraising efforts. Some public radio station managers openly accuse some members of Congress of sounding the death knell for public radio broadcasting in the US because of budget cuts to the Corporation for Public Broadcasting and National Public Radio.

With the budget cuts are steps taken by public radio stations toward sounding more "commercial" in their underwriting and program sponsorship efforts. Financial resources are limited. Public radio stations can either rely more on their sponsoring institution (which typically has gone through its own kind of "downsizing" and budget cuts) or become more dependent on listener donations and sponsor underwriting of its programs.

There is an increased need for public radio stations across the country to become even more creative in their efforts to raise money and continue to provide quality, alternative programming to their dedicated and loyal audiences. Now, more than ever before, those audiences are under increasing demands to have to contribute financial support for their programming or lose it.

## Summary

With the coming years there will be increasing demands by public radio stations for increased financial and voluntary support from within each station's local community. The practice of blatantly begging for money will increase at some public radio stations. Other stations will "disguise" their appeals by using promotional methods such as "Have A Heart For Public Radio" to reach their financial goals.

Noncommercial stations are facing growing challenges in terms of finding the financial resources to stay on the air and to continue to provide quality programming for their audiences. With less money available from sponsoring institutions and federal funding sources, there will continue to be increased reliance on their listeners for financial and voluntary support. Those stations that continue to grow and excel will most likely be those who are able to put a new, creative spin on the traditional promotional and marketing ideas.

Simply finding the money to start a new NCE radio station continues to be a major challenge. This is especially true when you consider the investment required in terms of new equipment, required equipment and the expert personnel needed to install the equipment, build the radio station, and then properly promote and market it once it is on the air. It is an even greater challenge to find the necessary funds to keep any kind of radio station on the air, whether it is an NCE station or a commercial enterprise.

Only through sound financial planning and creative ingenuity in recruiting volunteers and fundraising efforts will new stations survive their first few years on the air. This is why it is even more critical that you plan to determine who your target demographic is, that you target your programming to reach those people, and that you have a strategy for encouraging those listeners to help offset the rising costs of operation.

# Chapter 7 ▶ ▶

## EQUIPMENT REQUIREMENTS

---

### Chapter Outline

    A.  New or used (what needs to be purchased new versus what you can get away with buying used)

    B.  Technological considerations
CD/DAT/Hard Drive music playlist systems

    C.  Cost considerations (what can you afford) versus what you need

One of your largest investments in starting a student radio station is that of equipping your plant. This investment is actually twofold: monetary and a function of time. The amount of time you invest in researching both the kinds of equipment as well as in negotiating price can mean very large savings.

Investing in broadcast equipment is not unlike making other economic decisions. Typically you want the best quality at the lowest price. In our profession there are some manufacturers that have been in business for many years and who have solid reputations for producing quality equipment. These manufacturers were known for building equipment that would stand up to the most rigorous operating conditions and last for many years. These companies were also known to stock replacement parts and they had the ability to deliver overnight, sometimes within hours if the conditions warranted such immediate delivery. Some of this kind of equipment is still in circulation today and many commercial stations still use equipment from these manufacturers.

As mentioned several times in previous chapters, if you feel uneasy about starting a student radio station, by all means get advice from people who know about it. In terms of engineering, it is best to enlist the help of a broadcast engineer or join an organization which helps its members with engineering questions.

It is important to try to avoid relying totally on the advice of students who are taking engineering courses in your school unless they have engineering experience in broadcasting. Broadcast engineering is a specialty field and not every engineer is familiar with it. In some cases amateur radio operators can be very helpful, but typically new engineering students lack the familiarity with the specific needs of broadcast applications to adequately and knowledgeably inform you during this time when you need professional, experienced advice.

## Engineering Assistance

It is usually best to find an engineer who is currently working for a station, consulting in the field, or who has retired. There is no substitute for experience and you will find that good advice is invaluable when considering equipment purchases.

Technical assistance is usually available from the equipment manufacturers. It does help to have someone locally available to assist your station in times of need and to communicate with the techs as the equipment is repaired on your site.

Experienced broadcast engineers usually have their own opinions about what they prefer. This is only natural since they are the ones who will be called upon to repair it and being familiar with a given piece of equipment can be very advantageous. Every engineer is likely to have a favorite manufacturer for a given type of broadcast equipment, and usually their reasons go beyond mere personal preferences.

An example of just such a list of preferred used equipment types is included in the appendices of this book. It is offered not to plug particular manufacturers, but merely to serve as an informed example.

### New or Used?

The decision on whether you purchase new or used equipment should be based on your school or organization's economics. Quite often you can save 50% or more by purchasing good, quality used equipment from name-brand manufacturers. If you decide to invest in some used equipment, it is especially recommended that you have experienced engineering help available to you to make any necessary repairs or perform maintenance duties as required.

Other considerations in buying used equipment include the availability of parts and service from the company's technical department. Obviously, if the equipment is no longer manufactured (such as an RCA console, for example) you will also require good research skills to find replacement parts or to find an engineer who can fabricate necessary items or perform repairs.

An experienced broadcast engineer will probably be able to tell you whether you can go with new or used equipment in specific applications. An example is in the comparison of studio equipment to RF (radio frequency) or transmission equipment. It is much easier to repair or replace a cart machine in your studio than it is to make repairs on an antenna that is mounted 500 feet off the ground.

There are definitely some areas where the reliability of new equipment far outweighs the disadvantages of higher cost. Professionals who climb towers and have the necessary insurance can be expensive, and down time can be even more costly for your station. It is best to go with new equipment when considering an antenna and transmitter, as reliability of service is very important to your new station. Few listeners will sit and listen to static while you are trying to find parts for your transmitter.

When the time comes to replace your transmitter, it is also a good idea to keep your old unit as a backup if it has proven to be at all reliable. Some stations also have an auxiliary antenna or transmission line, should the main antenna or line become damaged.

If your organization or station is operating on a very tight budget, you might also consider used equipment donations. Typically a good place to start is with other commercial stations in your area which are seeking to replace their used equipment. Often this is "used up," but you might find donated equipment can be useful for parts. You may also want to consider running advertisements in some of the available trade magazines. Some magazines allow non-profit educational stations to advertise at no cost. Examples of publications where your station might seek donations are *Radio World* and the *Radio Shopper*. Please refer to the appendix of this book or this author for contact information for these publications.

There are many manufacturers and suppliers of new equipment. Make every effort to be sure that the equipment specifications match your station's particular needs. Again, seeking good technical advice can prove to be invaluable over time.

Another consideration in buying new equipment is cost. You will find a great deal of variation in prices from manufacturers. These costs often vary with the features and complexity of the particular kind of equipment. It should be noted that you don't always need to buy the unit with the most features, and optional features may lead to additional costs.

If your school or organization is considered to be not-for-profit, you would be wise to ask for a non-profit price. This may well result in huge savings.

Also, you are not required to pay tax on your purchases if your organization is legally registered with the Internal Revenue Service and your state corporation authority.

## Your Mission and Your Needs

What you need in terms of equipment should be determined by what you plan to accomplish with your station. If you plan to make it a training ground for broadcast students or plan to operate the station in conjunction with an established curricular program, you may need more studios to accommodate larger numbers of people. Many small market commercial radio stations find three studios quite satisfactory. The studio complement would include a main control room, a production studio and a newsroom. The quality of equipment would also probably follow that same prioritized order, with the best located in the control room and the "hand-me-downs" often finding their way into the newsroom studio.

In any case, unless you have a substantial budget and external support, you will most likely find that you can operate with somewhat less than "top-of-the-line" equipment. There are many decisions to be made. Do your research thoroughly and get as much professional technical advice as possible. Weigh all of the alternatives and make the most informed decision that you can based upon your financial situation and the availability of good local technical support.

### Your Station's Airsound: Audio Processing Units

Audio processing devices affect the overall "sound" of your station. Basically you need some method of controlling the audio signal to prevent overloading or overmodulating the transmitter. Typical installations include a form of AGC (automatic gain control which will expand soft passages and compress loud ones) as well as some kind of limiter to control peaks in modulation. If you are planning to make your FM NCE station stereophonic, you will need two of everything unless you invest in stereo units. If you use two AGC units, for example, it is best that you electronically "strap" them together with the interconnecting patch cord so that the units work in harmony together. Instructions for doing this are usually included in the equipment manual (if you are lucky enough to get one with the units). A knowledgeable radio engineer can help. Some audio processors combine various functions, such as compression and limiting (called comp/limiters). You may also choose to include some form of equalization, although some stations do not as it is an invitation for DJs to play with the units unless they are located in a secure area.

## Summary

If you are giving serious consideration to the idea of investing in good used, rather than new broadcast equipment, the best advice is to seek counsel from a qualified individual who has equipment experience and who knows where to find the best possible prices. They often have excellent reasons for preferring a particular manufacturer over another. Sometimes these reasons include a readily-available parts inventory.

Trust the expertise of a broadcast engineer. They usually really do know what they are doing, and they may save you appreciable dollars on used equipment purchases if you give them sufficient time to "shop around" and scan the trades.

Chapter 14 offers advice on free resources and can help steer you in some positive directions. Also, you may discover that an engineer who has been heavily involved in radio might be more helpful than a broadcast television engineer, and he/she might also be more available to help you. You may not have much choice, however, unless you live in a metropolitan market. Plan to "shop around" for engineering help. Although you can always pay for it, some radio engineers will volunteer their time or charge you a token sum for their time and effort if they believe in your cause and are experienced with/partial to noncommercial/student radio.

## Specific Radio Broadcast Equipment Requirements

### Typical Small Station Analog On-Air Studio Equipment Inventory

*Audio Control Console*
Tape equipment
Reel-to-reel(s)
Cartridge record unit
Cartridge playback unit(s)
Cassette recorder/player
Compact Disc Player(s)
Turntables

*Typical Small Station*
Analog Production Studio Equipment Inventory
Audio Production Console
Reel-to-reel(s)
(1) Cartridge record unit
Cartridge playback unit(s)
(1) Cassette recorder/player
Compact Disc Player(s)
Turntables
Bulk tape eraser*

*(this unit is usually kept in production studio and out of the on-air studio)

### Small Station Digital Production Studio Equipment Inventory

(Same as above, but console would be graphically represented on a computer monitor.)

Studio also includes Digital non-linear audio editing software and hardware, as well as one or more external hard disk drives for storage and retrieval of digitized audio.

*Transmitter/Associated Equipment*
Audio processing units (AGC, complimiter, etc.)
(a great deal of flexibility is permitted here as previously discussed)
Transmitter (AM or FM)
Solid State Exciter (if FM)

Transmission line (also called coaxial cable)
Cable connectors
Tower
Antenna (if FM)
Tower lights & obstruction painting*

*(if 200' or more above ground or if located in an approach pattern of a major airport)

# Appendix

## List of Used Equipment Preferences of One Engineer

*Studio Equipment*

| | |
|---|---|
| Consoles | Gates/Harris, LPB, Arrakis, Sparta |
| Cart Machines | ITC (International Tapetronics Corp.) Gates, Gates/Harris, Spotmaster |
| Reel-to-Reel(s) | Revox, Ampex, Otari |
| Microphones | ElectroVoice 635A, Shure SM57, etc. |
| CD Players | (can go consumer, such as Radio Shack) |

*RF/Transmitter Equipment*

| | |
|---|---|
| Transmitter | Gates (older) or Gates/Harris |
| FM antenna* | Gates (older) or Gates/Harris |
| | Phelps/Dodge, E.R.I. |

*Transmission Line*

| | |
|---|---|
| or coaxial cable & connectors | Andrew Heliax |

*Low-power FM antennae are available at a significant savings, but can only handle a few hundred watts per bay.

*Shively Labs (207) 647-3327 makes a new antenna at significant savings over other manufacturers.

## If Your Studio Is in a Different Location from that of Your Transmitter

Equalized Pair Telephone Line (prices vary dramatically can prove to be very expensive)

| | |
|---|---|
| Studio-Transmitter Microwave | LinkMosely or Marti |
| Studio-Transmitter Link Antennae (2) | Scala Parabolic or Mini-flector for short range |
| Coaxial cable lengths (2) | Andrew Heliax |

## If You Plan to Do Remote Broadcasting or ENG

| | |
|---|---|
| Remote Pickup Unit (RPU) (transmitter & receiver) | Marti |
| RPU antennae & coax | Marti |
| Cellular Telephone (depending on availability) | |
| Many available | |

# Chapter 8 ▶ ▶

## FINANCIAL RESOURCES AND CONSIDERATIONS

---

### Money

To properly start any kind of radio station takes money. This is certainly true for student or noncommercial educational radio stations. You might have heard once that a radio station license has been compared with a license to print money, but it really does take money to make money.

Legally a noncommercial radio station cannot air commercials. But there are noncommercial radio stations operated by students which are quite successful at selling underwriting announcements and sponsored (or underwritten) programs. These operators have thoroughly familiarized themselves with the rules governing underwriting and follow them to the letter. A discussion of these rules is contained in Chapter 12 of this book titled "Underwriting and Sales."

Despite these success stories, station start-up expenses can be sobering. Many of the costs are determined by your equipment needs, but they are also affected by the size, nature and power of your facility. Expenses can escalate if you are considering construction of a directional AM station over a non-directional, or a high-power FM over a low-wattage facility.

A first step is to determine a budget. Establish early in the game just how much your organization has available and begin planning how you will invest it. Just as a good engineer can give you excellent advice when it comes to equipping your station, a good financial consultant can offer a great deal of helpful assistance in the area of economics and budgeting.

A great number of questions must be answered before you begin. One of the first is based on your estimated construction expenses and first year operating costs: do you have enough financial support and cash on hand to start this venture? If not, how do you possibly plan to raise the capital necessary to accomplish your goals?

## Setting a Budget

Companies with an eye are able to project how much money it costs them to produce a product.

Even if your school, college, university or organization picks up the tab for you, you still need to think in terms of budgeting your money.

In these tight economic times it is much harder than it used to be for an institution to pay all of the bills associated with running a radio station, student or otherwise. There are always going to be operating expenses, even if your station is tax exempt and the building your station occupies is paid for free and clear.

## Typical Budgetary Considerations/Line Items

There is a myriad of different situations and no one scenario or budget can be suggested for everyone. However, here is a list of possible expenses that you should consider and investigate. This should assist you as you prepare the budget for your new station:

### New NCE Station Budget Line Item Suggestions

1) Construction costs—Materials and labor
2) Equipment purchases
3) Promotion
4) Utilities
5) Telephone
6) Property Insurance
7) Taxes/liability insurance
8) Rent
9) Consulting fees
10) Salaries/wages/commissions on underwriting sales
11) Printing costs (flyers, program schedule or guide, etc.)
12) Advertising
13) Music Licensing (ASCAP, BMI, SESAC)
14) Postage
15) Depreciation on any new equipment

## Expenditures

The smart NCE station operator will determine what the total of these expenses is. Each line item on your expense sheet should be carefully evaluated by your supervisory board, or your station's financial officer. Is each expense justified and reasonable, or are you spending too much in any given area or in multiple areas? Are there certain areas where your expenses can be reduced or eliminated? How do your expenses this year compare to those of previous years? Have there been major changes in your station's operation that have influenced these changes? Do you think you could project your expenditures for the upcoming fiscal year?

These are all excellent questions that you should consider answering when you review your expenditure sheets. Any money that your station can save in terms of reducing operating costs is money that you have saved.

## Income

The next step is to list your station's income from any source. Does your station sell underwriting or sponsorship of programs? Are you making any money in your efforts? How does your income compare with that from previous years? Do you think you could project your income for the upcoming fiscal year? Would it result in positive cash flow for your station, but yield a net loss?

Some NCE stations are owned by a non-profit corporation. This is particularly true of community stations. As non-profit corporations, they must show a net loss at the end of their fiscal year. Some of these operations are known as 501(c)3 corporations and must complete and file this form with the Internal Revenue Service. It is important that you plan to work with a qualified accountant or firm, preferably a Certified Public Accountant (or CPA).

If your station licensee will be a board of trustees or the regents of a university, for example, then the institutional accounting department may administrate the station's books.

## P&L Sheet

In the world of accounting and commercial stations, accountants maintain a profit/loss (P & L) sheet. This tool helps to balance where profits and losses are occurring. It is then quite easy to determine where the money flow is going, and in which direction. Your accountant will likely require that you set your books up in a specific manner and that you maintain detailed records of all of your financial transactions.

## Balance Sheet

Similar to the P&L, this helps station management personnel keep a system of checks and balances on where monies are flowing using standard accounting methods and line-item entries. This is also very useful when comparing performance on a year-to-date basis.

## The Bottom Line

The bottom line for your new station will probably involve a lot of red ink. You will probably operate your station at a loss for several months. In some cases NCE FM stations must expect to make payments on loans written by their sponsoring institution for power upgrades and new station additions.

### Sound Advice for New NCE Student
### Radio Stations on a Budget

1) Operate your new station in as lean a manner as possible.
2) Utilize volunteers.
3) Keep paid staff positions to a minimum.
4) Plan a projected monthly and annual budget for the first 12 months of operation.
5) Identify and list potential income streams.
6) Have a clear, written policy and understanding with your station's licensee regarding whether underwriting/program sponsorships will be sold or not.
7) Brainstorm possible fundraising events (if your school permits it).
8) Form a committee to investigate revenue-generating possibilities and to formulate new ideas.
9) Place an individual in charge of the underwriting department.
10) Place a committee or individual in charge of overseeing the station budget, income successes, etc.

## Where to Find Money

This is always a popular but crucial consideration. In the realm of commercial radio station business proposals, entrepreneurs who have a well established track record can approach a bank or other lending institution and request a loan. If you are building a new radio station at your school there are several funding possibilities. You must, however, carefully consider all of the future ramifications which may come with the financial support you may accept.

Typically a college or university is in a position to finance such a new station. If it is a student station and you are particularly interested in maintaining student control, you might seek financial backing from your student government.

Another option is your school's broadcasting or journalism department. However, in some situations these entities may consider the radio station project as an excellent outlet for student lab projects. In these cases, if the funding is made available by a particular department, you can probably expect them to have extensive control over how the station is operated and its mission.

If your station will be affiliated with a high school, junior high school, or other public or private school institution, where you go for financial support may determine how your station is controlled in the future. If you go to the local school board, school district or superintendent and request financial backing, you can probably expect that they will want some control over how the station is used, who can be on it, and the station's mission. There is always the possibility that the superintendent or school board may wish to use the station as a public relations tool to build credibility with taxpayers and voters in the community, for example. It is important to note that, just because an NCE is student-run or student-operated does not mean that it is student-controlled.

Typically, whoever controls the purse strings has control over the station and has final authority over its operation. Therefore, it is important that you carefully consider the issue of where the money is coming from and how much control this source will inevitably have in overseeing your operation.

## Summary

Running any kind of radio station is a challenge. Running an NCE station on a budget in this economy can be frustrating. The keys to success are planning, budgeting, supervising, and making main-course adjustments while underway. Plan to use your resources wisely and prudently, and have a good accountant available for advice you may need.

Remember, money is power. Invariably, wherever your initial funding comes from will also probably determine who has the major voice in how your station is operated.

# Chapter 9 ▶ ▶

## PERSONNEL RESOURCES

---

*(Where to find and keep qualified people)*

### Personnel Requirements in Commercial Operations

Your air sound is the product for attracting listeners. In addition to music, people (voices) are a major factor in your programming. Ever since the years of personality radio in the 50s and 60s, listeners have been attracted to the voices and style of particular talent. From Larry King to Howard Stern, individuals are the commodity that attract listeners.

It can be very difficult and costly to recruit, attract and retain talented air personalities. In some cases well-known air personalities have actually exceeded the income of some executives at particular radio stations. A major portion of the payroll of many small market and medium market radio stations is comprised of wages and salaries (including benefit packages and payroll taxes). Personnel are what makes the wheels of your radio station turn. You can still have listeners without having a local on-air staff, but it becomes very difficult and challenging to respond to community needs.

Some small radio stations have decided to carry satellite programming 100% of the time they are on the air, while some others elect to simply air a live morning show or morning team, then go satellite for the remainder of the day and night. This arrangement can save stations a lot of money, but can make them totally unresponsive to the local listening audience. In one such station, after the live morning show from 6am until 9am, the station doors are locked until the next morning. Ten years ago this would have been a violation of the FCC rules. With the advent of more dependable technology in terms of transmitters and monitoring equipment, however, the new rules permit "walk-away" operations. Dial-up remote transmitter monitoring systems and automatic devices which can record all operating parameters of a transmitter have brought the age of "walk-away" operation into reality.

In addition to the technical aspects of automating the programming and transmitter monitoring areas, some frugal commercial radio station managers have also managed to minimize the number of staff and support personnel in their other operations. One example of this is the sales staff that is partially or totally comprised of individuals who are selling advertising and compensated for their efforts on commission only. This "commission on collections" arrangement means the sales staff is paid only after the station is paid.

Through the use of computerized voice mail, e-mail, and the world wide web, communicating with stations has dramatically improved. By incorporating analog or computerized answering devices, some small commercial radio stations have managed to eliminate at least one additional staff person: the telephone receptionist. In other cases commercial station operators have often combined the duties of the telephone receptionist with those of a regular receptionist (to greet station visitors in person) and traffic director. Often this individual would juggle the duties of preparing and typing the next day's program log, answering the telephone, and greeting people.

Satellite programming, the automated DJ, and the computerized or dial-up transmitter meter reader, and commission-on-collection sales staff, have enabled some small and medium-market commercial radio station managers to cut their payroll.

## Personnel Requirements in NCE Operations

Student radio stations are likely to be blessed with a plethora of help. Their "sponsoring" school or institution is often in a position to provide many of the operational items a station needs. In the case of a school, there is usually a large number of students interested in becoming involved with the student radio station.

Although volunteer help is seldom lacking in a student radio station environment, one or more positions may be paid. This may include the station manager or heads of some departments, including station underwriting, programming, sports, news, or engineering. Paid positions in student radio stations are dependent on the structure of the operation, the management philosophy, and the educational institution.

One of the greatest assets of student radio stations is the seemingly unlimited amount of creativity and ambition associated with such operations. It is the desire of the participants that propels student radio stations forward, upward and onward. Many listeners enjoy hearing alternative forms of programming, new music that is not available elsewhere on the dial, and announcers who really know their music and who aren't afraid to try new things and ideas on the air.

This "uniqueness" of student radio is largely a function of the fact that their operations are often "unfettered" by the necessity of dependence on listeners and receiving advertising dollars. Usually their school provides the plant and the operating funds, thus relieving student station managers of that heavy burden of responsibility. They are free to concentrate on programming ideas and to get creative with their operations. This is not to say that all commercial radio stations are not creative, but there are certainly some very creatively-programmed and operated student operations on the air across the country.

One of the most rewarding challenges in student radio comes from successfully and effectively managing this pool of readily-available talent. If commercial radio station managers are concerned about personnel turn-over, imagine the headache that is faced by student radio station managers. Every year, every semester there are students joining and leaving the staff.

This turn-around in personnel creates a whole new set of problems that must be faced by the student station manager. These problems sometimes surface in the form of theft of station CDs and/or records, irresponsibility in the form of talent not showing up for air-shifts, and profanity/obscenity on the air. In a "free-form" station the announcer may believe that anything goes, yet the school's administration and/or the FCC have other philosophies. An entirely different set of challenges and situations can be found in student radio stations compared to their commercial counterparts.

Quite often the only "stick" that can be held over student operators is the threat of suspension or termination, and being let go from a non-paying job may not mean a lot to some students. If the student radio station has established a good reputation and it is considered to be a privilege to work there, then the "carrot and the stick" philosophy can be effective.

Some colleges and universities have fully spelled-out regulations for student conduct. If a policy is established early in a station's development that misconduct at the station will fall under the school's rules for student conduct, it can provide a motivation for students to follow station rules. There is a big difference between being kicked off the air or terminated from a non-paying position and being kicked out of school. If you are considering the implementation of such a policy, you will obviously have to earn the support of your school's administration and work closely with them to have just such a policy established for your station.

## Community Personnel Resources

It is important not to overlook personnel resources that may be available in your own community. For the purposes of this discussion, your community should include the people within the walls of your building.

The most obvious personnel resource is student volunteers. However, do not overlook the opportunity to involve faculty, staff and administrative members in your school. Some stations have successfully involved their school presidents as well as faculty and student volunteers. Students get a real "hoot" when they hear their teachers pulling an airshift and are often really surprised just how good their teachers are on the air. This can be equally rewarding and fun for the teachers, staff members and administrators who get involved with the station. They have fun and discover it is another wonderful way to become involved with their students.

Of course, some faculty, staff and administrators may not want to become involved, but they will probably appreciate an invitation. If you can envision creative opportunities for involvement, you will have more successful results in your recruitment of faculty, staff and administrative volunteers. This could include weekly pre-recorded or live programs such as "Five minutes with our Science teacher, Mrs. Roberts," "The Professor's Corner" or even "The President's Corner." Such programs can get really popular if you incorporate some kind of interactive element, such as a call-in opportunity, for listeners. Administrators are likely to realize the value of such a program as an opportunity to present their views. Their active involvement should be welcomed and encouraged, but it is important that they remain as autonomous as possible. There have been cases where struggling (and sometimes not-so-struggling) student stations have been used by school administrators or broadcasting departments for student laboratories, or as public relations outlets for the school, college or university.

On a positive note, there is an added benefit to recruiting and actively involving teachers and administrators. Active involvement by teachers, staffers and administrators in the operation of your station may well result in their developing a good feeling about the station and your programming, which in turn may open doors to new resources.

Starting a new student radio station offers a wonderful opportunity to be really creative. Music programs, live (over-the-telephone) or taped remote broadcasts of school events will really attract listener involvement and awareness of your station. With a new station there are all kinds of new opportunities.

Tap the resources of your listeners and student body. Your creative promotions department can design station logo contests, for example. If the promotion is designed with input from your underwriting department, you probably can find a local business which will help support your effort by donating a prize or gift certificate to be awarded to the winner. Usually local business operators view these kinds of opportunities as an avenue to reach potentially new customers and to reinforce public awareness of their business. Be cautious that the "thank you" recognition announcement you broadcast is not a commercial in disguise. You may want to review the FCC's rules regarding underwriting contained in chapter twelve of this book if you are considering underwriting as a potential revenue source.

Another valuable personnel resource can be community and business leaders. Extend your station's presence beyond the school's domain, and you may gain listeners, community support, underwriters, and a genuine community presence. Leaders such as the mayor, city council members, police, fire officials, and the like can also prove to be excellent guests and supporters. You can provide a valuable service to your audience if you coordinate appropriate guests during certain weeks, such as fire prevention week, crime prevention week, and such.

Remember that your new station can be so much more than just a juke box or training ground. Take advantage of the uniqueness of student radio and explore all of the possibilities.

## Volunteer Recruitment

You will find that it is relatively easy to recruit student volunteers if you promote frequently. Successful student radio stations are very active in recruiting, training and retaining their staff members. Promotion is critical to the success of a new student radio station in terms of both eliciting student volunteers and acquiring new listeners.

Some stations will broadcast live remote broadcasts during registration periods at the beginning of each semester. This is an excellent way for your new station to promote itself. Student stations will often distribute bumper stickers, station T-shirts, flyers, newsletters, station application forms, and other station propaganda to attract new volunteers as well as new listeners.

The number of volunteers a given station (or student organization) receives at the beginning of a new semester is directly proportional to the amount of effort and the perception of fun that they convey to potential volunteers. If they can convince interested students that they will have an opportunity to actively participate in the station, they will be successful in their recruitment efforts.

## The Station Orientation Meeting

A station new-member orientation meeting can advance this thought and solidify this perception in the minds of new volunteers. These kinds of orientation meetings should be carefully planned, professionally presented, efficiently organized and limited to one hour. Time should also be available near the end (or after specific presentations) for questions.

Orientation meetings should include presentations from all department heads, or at least all areas of station operation. Often students want to be on the air and the programming department will get the majority of volunteers. If your proposed station plans include an underwriting department and a sports or news department, you should also make efforts to involve new volunteers in these areas.

Equally important to the success of your operation is orienting new volunteers to all opportunities at the station. Even student stations need support staff, including people that can answer telephones, come up with promotional ideas, help with the equipment, design logos, and such. There are many more opportunities available to students than the obvious one of being on the air.

## Volunteer Training

In planning your new student station, you might consider the establishment of a training program. If your school is lucky enough to already have a broadcasting program, cable-FM, PA system or carrier current station and you are getting an over-the-air station, do not abandon your other operation. Instead, use it as a training ground for new volunteers. In this way you can orient them to your operation in a "hands-on" environment while having an opportunity to observe them. This will allow you to determine whether they are reliable, responsible, and mature.

A PA system radio station is like having a second radio station, and it can be a great training ground. It can also keep volunteer morale and belief in the program alive while they sit out the long wait for the FCC to process your application for a construction permit.

## Coordinating Volunteer Efforts

In the orientation and training process of volunteers (both student and non-student), it is important to develop a program of retention and supervision. Some student stations will not have the large numbers of volunteers that are enjoyed by their larger counterparts. This can be particularly true of smaller schools including high schools and community colleges. Do not let this fact dishearten you. There are many small, yet very successful stations that remain locally active and responsive to their listenership while maintaining small volunteer staff sizes.

The level of success in coordinating your volunteer staff may be directly proportional to your own managerial skills and ability to delegate authority. In stations that have large numbers of student volunteers, you may find it more efficient to select (or elect, depending on your management structure) department heads. Even though these may be unpaid positions, especially at first, the individuals in these positions will be motivated by ego, the additional responsibility, and the excitement of having an even more active role and voice in the operation of the station.

It is important that every person involved with your new station understand the big picture. This includes the responsibilities of others involved in other departments or areas. Of equal importance is the establishment of a customized training program for each particular position. This can be coordinated through the department head.

## Orientation Handbook/Policy Handbook/Job Descriptions

An orientation handbook and a policy handbook will make coordination of your volunteers more efficient. The effort you invest in creation of these documents will help ensure an efficient, well informed volunteer staff.

Job descriptions for each volunteer position will also help orient your new volunteers and lessen the station manager's workload, as well as that of department heads. Many questions can be answered in advance for new staff members if the handbook is planned and written in a well-organized manner. You should also be open-minded to suggestions for improvement, including additions and rewrites of these documents.

## Staff Meetings

In even in the smallest student radio station it is important to maintain good communication with your volunteers. Air staff meetings, underwriting meetings, department head meetings and/or general staff meetings, help everyone to know what is happening at the station. These meetings are excellent opportunities for planning, implementing, goal setting and brainstorming.

Meetings should never exceed one hour in length. When possible, try to have the meeting last less than an hour. Have an agenda and follow it, allocating a time limit for each item to be discussed during the meeting. If the time limit is exceeded, determine just how much more time will be allocated to discussion. If you must extend the time again, you may choose instead to table the item until the next meeting.

General staff meetings should be held on a weekly basis. Less frequent meetings are not sufficient to keep everyone appraised of station happenings. Department head meetings should also be held weekly, as well as department meetings. Although this may sound very professional and too organized for student media, you will find it will help you to coordinate the staff and department heads more efficiently. The meetings should be held, even if they only last a few minutes each.

## Summary

Here are some key thoughts to remember: have a mission for your station, set goals for your staff, promote your station, actively recruit new volunteers, involve faculty, staff and administrators in the operation of your station at as many levels as you can, plan, organize and hold staff meetings consistently, and widen your horizons: open your mind to new ideas from your staff, for collectively they represent your success or failure. Your staff is the power behind the new station: capitalize on their energy, excitement and enthusiasm. It is important that you turn your new station in to an idea factory, and involve as many people as you possibly can.

## Conclusion

Your new station's staff is more than just a collection of ambitious, energetic people who want to form a station community. They are a wellspring of creative ideas and promotions. Keep an open mind and brainstorm about ideas and solutions to challenges. Strive to treat everyone with respect, and encourage your staff to do the same for each other. You are encouraged to do more with your station than just play music: you are challenged to serve your listeners, and to provide information as well as entertainment.

And above all, make your station a fun place to work and learn.

# Chapter 10 ▶ ▶

## EQUIPMENT RESOURCES ON A TIGHT BUDGET

### New and Used Equipment Resources

One of the largest initial expenditures your new student radio station will have concerns equipment. You will need to decide whether to purchase new or used equipment. This decision will probably be largely based on your budget and the availability of local broadcast engineering assistance.

If your station is starting with a limited budget, an option that you may wish to consider is to purchase new equipment in critical areas, and go with used equipment in other areas which you might later replace or upgrade as funds become available. Critical equipment might be considered to be your antenna (if FM), transmission line, tower and transmitter. Because dependability is vital, you should invest in new equipment here if your budget will permit it.

If your budget is extremely limited and you are applying for an FM station, you may still seriously consider investing in a new antenna. The antenna is something that cannot easily be replaced, as tower crews can be expensive and not always readily available. Weather can have a great deal to do with replacing an antenna, as well, and may result in your new station being off the air for several days while a new antenna is ordered, received, and installed.

Conversely, a transmitter can typically be replaced in one night with little or no down time for your station. A good FM antenna can enhance your signal coverage and a good match with your transmitter will assure maximum efficiency meaning maximum wattage radiated. Investing in a new FM antenna should be a very high priority.

### Locating Good Used Radio Broadcast Equipment

Some people have a natural talent for finding good used equipment at reasonable prices. Actually, there is no real secret to locating good used broadcast equipment. Much of the success comes from having a lot of industry contacts and networking. Making friends, subscribing to one or more ListServes appropriate to your interest area(s), and reading the trade magazines are the keys to success here.

If you do not want to read several trade magazine classified advertising sections each month, receive two dozen electronic mail messages daily, or spend a few hours a week on long distance telephone calls, you might not be cut out for this kind of activity. If you are in need of quality used broadcast equipment, you should contact someone who does have the skills and contacts that you need.

Some broadcast equipment suppliers do offer used broadcast equipment. In addition, some companies have actually specialized in used equipment. In these cases, it is best to beware. Sometimes such companies take advantage of a neophyte's lack of negotiation skills or sense of what the equipment is worth.

Some used broadcast equipment vendors specialize in specific types of equipment. Some of them focus their energies on used transmitters, and still others concentrate on selling new and used audio or recording industry equipment. Sometimes these companies have what you want. In other cases you might wish to engage the services of someone who has the skills necessary to be a "good used equipment broker" who may find what you need at a reasonable price.

Some people are very adept at solving mysteries, finding used equipment, and doing all of these things at a very low price. Some of the following used equipment sources may prove to be valuable: *Broadcasting & Cable, Radio World,* and *The Radio Shopper.*

### Broadcasting & Cable

This trade magazine is available from Cahners Publishing, Inc. on a weekly basis. Of particular interest to used equipment buyers and to students looking for jobs is the classified section. Advertisements are divided according to radio and television, and further subdivided by category. One such category of interest to equipment buyers is "For Sale Equipment," typically located near the last page of the publication. You can contact *Broadcasting & Cable* at their web site: http://broadcastingcable.com.

### Radio World

This publication comes out bi-weekly. Subscription is free to industry professionals and stations. *Radio World* includes an extensive classified advertising section called "The Broadcast Equipment Exchange." Each category is subdivided into "Want to Buy" and "Want to Sell" areas.

Sample categories include cart machines, computers, consoles, microphones, monitors, receivers and transceivers, recorders, test equipment, turntables, and transmitters. There are also sections on "help wanted" and "positions wanted." Contact them for more information at: *Radio World,* P.O. Box 1214, Falls Church, VA 22041-0214, voice: (703) 998-7600, *Radio World's* Fax number is: (703) 820-3245.

*The Radio Shopper*

This monthly publication is similar to *Radio World* in that it offers a fine variety of categorized advertisements for used broadcast equipment. It describes itself as "Radio's Equipment Forum," and is published by Radio Press Group, Inc., with Raymond C. Topp presiding as publisher. The address is 511 18th Street, SE, Rochester, MN 55904. Phone 507.280-9668. Fax 507.280.9143, their e-mail address is radio@broadcast.net and you can visit their publication and request information at their website address: http://www.radioshopper.com.

It should be noted that there are numerous other publications available to help you in your research. These three publications seem to offer the most help to shoppers for used equipment.

In addition, the value of word-of-mouth cannot be emphasized enough in this discussion. Simply passing the word around that you need a used Burk or Gentner remote control will almost assuredly get you some kind of response sooner or later. The real key is time. If you can wait until what you need is located, you are most likely going to save your station some money. If, on the other hand, you are in a hurry, you should be prepared to pay more.

Do not overlook commercial and non-commercial broadcast stations in your area. Sometimes all it takes is someone to ask, and these stations can find the most amazing things in their attics and garages.

And certainly include television stations on your "equipment hit list." Even though you are looking for radio equipment, television stations do have audio and at times are willing to donate some of their old audio equipment such as microphones, mike stands, consoles, and even audio processors to help a new NCE radio station get on the air. It is certainly worth your effort to contact them. Another source is the National Weather Service. One NCE station turned this author on to the NWS as an equipment source and as a result had many like-new tape cartridges and cart machines donated to the station when the NWS dumped their old analog equipment for digital operating systems. The cart machines were top-of-the-line in their day and were in operational condition.

There is also a lot to be said for the wisdom of age and experience. You can probably find the equipment you want, but it takes an experienced radio broadcast engineer who has worked with a lot of this equipment to know the right questions to ask about any given piece of equipment. Again, you should try to find someone with this kind of experience who is willing to help you in your efforts. Brush up on your networking skills and spread the word around about what kind of equipment your station needs and how much you are willing to pay. In some cases you can request that equipment be donated, but you may not always get equipment that is in top-notch condition.

Many familiar mottoes come in to play in this game. Of particular note is the one that says "Let the buyer beware." Just because a piece of equipment is the one that you want does not mean that it is in good condition, or that you can still find replacement parts for it, or that you can even find an manual for it or someone who is familiar enough with it to repair it.

As with other areas in this book, the advice "Do not overspend your budget" still holds true. It is easy to overspend in equipping your station. It is natural that you want it to sound the best that it can.

Conversely, as with a very expensive stereo system, sometimes a few thousand more dollars for a piece of equipment will not necessarily pay off for your listening audience. Take care and invest wisely in your new and/or used radio broadcast equipment shopping endeavors.

## Summary

Read the trades that are mentioned here relentlessly. Use your phone to check out leads. Be selective in what you inquire about through your searches. Network with other students and broadcasters. Inquire and ask for equipment donations from local and area radio and television stations. This can be a particularly effective source for used equipment since you are starting a new station and these broadcasters know that you need all the help you can get.

Above all, enlist the help of a radio broadcast engineer who likes students, student radio stations, and is more than willing to help you in your efforts. Most radio broadcast engineers just love equipment. They especially like it if it works and if they are not going to be responsible for daily maintenance or repair on that equipment.

After you have successfully enlisted this help, explain to the engineer what you are trying to do, what you would like to do, and ask for an opinion on just what equipment you will need to accomplish your goals. If you are on a limited budget, be sure to make this clear. If you are on an extremely tight budget, explain that you need to get as much as you can via the donation method. Also mention to as many people as you can that you are looking for used broadcast equipment donations and see what you get.

Try to work with a broadcast, or preferably a radio broadcast engineer, and not just an engineering student in your school. There is a big difference: just because someone is studying engineering, or even electrical engineering, does not mean they can properly install or repair radio broadcast equipment.

Many radio broadcast engineers tend to pride themselves on being a little independent, a little different, and are usually quite set in their ways. You may find it interesting that sometimes an engineer will come into a station, inspect a predecessor's work, and then say it is not satisfactory. At this point they will tear it all out and replace it all. The lay person will look at it and not see or hear any difference, but radio broadcast engineers like their own way of doing things. Some radio broadcast engineers seem to dislike interference and managers who hang out over their shoulder. Some may prefer to work alone.

Do not think that this behavior is unusual or unnatural for engineers: this is just the way that some engineers behave. Many of them are typically the most friendly and willing people you will ever meet. Some people believe radio broadcast engineers live in a world all to themselves, but they bear the burden of a lot of responsibility and they are often on call like a doctor: twenty-four hours a day.

A final word of caution: never discard used radio broadcast equipment. As Murphy's Law often works overtime, you may be surprised that someday you may need the very piece of equipment that you threw away. In fact, it may just include the one part you are unable to find in a hurry to get your transmitter back on the air. Cannibalization in radio broadcast equipment is a very common occurrence in engineering shops and circles around the country.

You might also be able to trade this seemingly worthless piece of equipment to someone at another station for something that would be valuable and beneficial to your station. Save and swap your used radio broadcast equipment, and make sure you have backups on hand for your most critical devices, such as your transmitter. Good luck always in your used equipment scrounging efforts, as well as scrounging for a good, knowledgeable radio broadcast engineer.

# Chapter 11 ▶ ▶

## ESTABLISHING STATION POLICIES AND PROCEDURES

---

### Policy and Orientation Handbook[s]

No one likes rules that are made up as the game progresses. Likewise, everyone that works at a station needs to know what the rules are, so that they can understand how the game is played. It is easier to understand right from wrong if the rules exist, are known by everyone, and are available to everyone in written form.

This written form usually takes place as a rulebook or policy handbook at many radio stations. Station policy handbooks set into writing the policies and rules that govern the operation of the facility. The primary intent of a policy handbook is to clearly and concisely communicate to employees and volunteers exactly what is expected of them, and which kinds of behavior are deemed to be unacceptable.

What follows is a very simple employee's handbook which includes a section on orientation. Although it is very simplistic and not very complete, it is included here to give you a practical sample how such a document is written and appears. It is no way intended to be complete, or to be adopted by your station. If it gives you some ideas of areas which could be included in your policy handbook, please feel free to borrow them to incorporate into your efforts.

*Some Suggested Elements for Inclusion in Handbook*

1) Attitude
2) Behavior
3) Professional Expectations
4) Dress Code
5) Misconduct Policy/Termination Procedures
6) Vacation/Holiday Policy
7) Sick Leave Policy
8) Telephone Etiquette

# Encyclopedia International Employee's Handbook

## Welcome

Welcome to Encyclopedia International. This training notebook is designed to give you a general introduction to the company. We hope it will answer many of your questions, but please don't hesitate to ask about anything not covered.

## Orientation

Your group assistant will enroll you in an employee orientation session on the first Monday after you begin work.

At orientation, you'll learn some of the general policies of EI. You'll also be asked to complete some forms for the Benefits and Insurance office. And you will be introduced to your guide for the day—most often a member of your department. All new employees receive a special gift during orientation as well, so be sure to attend!

### Checklist for the First Day

During your first day, you will meet the people you'll be working with, and become acquainted with the EI facilities and procedures. Your guide will help you complete the checklist below.

### Meeting Your Manager

During your first day at work, you'll get to meet your group manager. You'll want to get to know this person. He or she is the person who keeps things moving smoothly, coordinating projects and people assigned to your group.

### Tour

The EI campus is proud to boast five new buildings in addition to the original offices constructed in 1981. In addition to the office buildings, there now are an on-site printing facility and shipping area. Employee facilities now include two dining halls and a day care center. During your first day, you'll get to see EI's excellent facilities for yourself.

### Getting Equipment

At EI, we try to provide the equipment you need without delay. During your first day, the group assistant will show you your office and help you find the equipment and supplies you will need. If we overlook anything, don't hesitate to ask your group assistant for what you need.

## Questions About

This section of the handbook covers some of the subjects new employees frequently have questions about. The sections are brief, explaining generalities. You can find more information on all of these topics in the Training Notebook.

### Office Supplies

Your group assistant can help you get any office supplies that you need. Most are available on request, but there may be a delay of up to two days for special requests such as hanging file folders. If you need office equipment or furniture, talk with your group manager about putting in an order. EI wants to make your work life as comfortable as possible.

### Insurance and Benefits

EI offers a good package of medical and dental insurance. In addition to stock options, EI offers a savings plan that matches up to 7 percent of an employee's contributions after the first six months of employment.

### Vacation, Personal Leave, and Sick Leave

All employees are entitled to two weeks of paid vacation each year during the first three years employment. This increases to three weeks per year after three years, and to four weeks per year after nine years.

Employees also are entitled to one-half day of paid sick leave per month. Any unused sick leave at the end of a year is kept on account for the coming year.

In addition, EI offers an excellent maternal and paternal leave plan, and a day care center in Building 5 is open to employees' infants, toddlers, and preschoolers. Enrollment is on a first-come, first-served basis. For details, see your group assistant.

## Author's Policy/Orientation Handbook

What follows is the first draft of a policy handbook written by the author. This policy handbook is specifically written to address issues which may arise in the operation of an NCE station.

This handbook was conceived and written both as a sample for this book and with an eye toward the future. This "dream" represents one of the author's goals of starting a chain of NCE AM & FM student radio stations associated with a school of broadcasting. It is planned to also have an AM commercial station license to give students the opportunity to experience actual commercial station operation and sales.

The stations are licensed to The Great Lakes Broadcast Academy, Inc., a non-profit corporation licensed by the State of Michigan. The school will operate under the corporate name of The Great Lakes School of Broadcasting, Inc., and shall be duly licensed by the Michigan

State Board of Education and fully bonded in accordance with the policies of the Michigan State Board of Education in Lansing, MI.

It should be noted that this is a first draft and is incomplete. However, it is included here to give you an idea of what kinds of issues should be addressed and included in such a policy/orientation handbook. The author decided to combine the Policy Handbook and Orientation Handbook for the ease and simplicity of the project, and to reduce the element of confusion for the potential user. Again, this author must stress that it is not complete. In fact, your policy handbook should always be open to amendments, changes, and additions as time makes the need clear to you.

## Handbook Examples

### Hypothetical Example

Q-92 FM

AM-71

Great Lakes School of Broadcasting, Inc.

POLICY/ORIENTATION HANDBOOK

Drafted: 1996

Authored by:
James J. McCluskey, Chairman of the Board and Mary Sue McCluskey, Co-Chair of the Board and Co-Station Managers

---

*Contents*

I. Mission Statement of the Stations—Purpose

   A. New Staff Station Orientation

   B.  Supervisory Staff

   C. Warning Procedures

II. Expectations

   A. Professional Conduct

      1. Dress Code

      2. Telephone & Interpersonal Etiquette & Protocol

      3. Written memo & message format

      4. Conduct outside station, in public

   B.  Grounds for Misconduct

   C. Warning Procedures-Dismissal from staff

III. Staff Evaluations

## Mission Statement of the Station(s) and Purpose

First and foremost, the station(s) hereafter known as Q-92FM and AM-71 are licensed to serve in the public interest as a public trustee. As the licenses are held entrusted to the Board of each corporation, the Chairman and Co-Chairman of the board are directly responsible for overseeing their operation on a daily basis. It will be presumed that every minute of every hour the station(s) are in operation shall be as professionally run as humanly possible. Quality programming and serving our audience with the highest level of professionalism shall be the first goal of the operation of each station.

Additionally, successful and dependable participation in several facets of the operation of each station on a daily basis shall be considered to be a laboratory experience for every student as part of the graduation requirements of the Great Lakes School of Broadcasting.

The AM station, licensed on a frequency of 710 kHz. shall be considered to be the first training ground for all student operators. Since hands-on experience is required for every student on each station and course credit is awarded for participation in several facets of station operation, no financial remuneration for production, announcing or talent services shall be provided.

## Part One: Orientation to the Stations AM-71

### AM-71 Structure

AM-71 is licensed as a commercial AM Daytime-only operation. The license is held and the station is operated by a non-profit corporation legally registered with the State of Michigan. Voting stockholders and primary corporate officers include James and Mary Sue McCluskey, who are likewise responsible for the daily operation, management and supervision of the station, unless another individual is so designated by them on a temporary part-time or full-time temporary or permanent basis. Corporate control of the station shall be made by the McCluskeys or, in the event of their deaths, their heirs or assigns.

### Defining the Community

The station is licensed to operate from local sunrise to local sunset only on a frequency of 710 kHz. with a power of 500 watts, with the city of license as Stanwood, Michigan. Sign-on and sign-off times are dictated by the FCC as part of the instrument of authorization and license. Other cities identified in this stations legal identification shall also be Big Rapids, Reed City, Morley and Lakeview. The station also serves the communities of Rogers Heights, White Cloud, Newaygo, Greenville, Mecosta, Edmore and all areas within a radius of about 50 miles.

### Sale of Underwriting and Program Sponsorships

As a commercial entity, it is permitted to financially support the station by the sale of commercial time to area businesses, as well as other promotional and marketing ideas. It shall be the policy of this station to solicit underwriting, donations or public support from listeners. Any such solicitation MUST be under the direct supervision of an instructor and the student must also be currently enrolled or have successfully completed the sales and underwriting course.

### Student Conduct

Although this stations coverage area may be somewhat limited as compared to other area stations, including our plans to upgrade the sister station, Q-92FM, personnel working on this station *MUST* conduct themselves in a professional manner at all times. This goes beyond simply their on-air and production work and extends to in-house as well as public appearances and conduct in their everyday lives. It should go without saying that everything every student and faculty member does and says is a reflection on the integrity of this station and school, which is always being closely scrutinized by the public, the Michigan State Department of Education, the FCC, and numerous other national, state and local organizations of which the school and the stations are members. It is mandatory that all station personnel dress, speak, and act as professionals at all times. Any digression from this mandate shall be considered as an act of misconduct, and shall fall under the penalties discussed later in this policy handbook for such violations.

## Community Service

As a commercially licensed station, AM-71 is legally allowed to sell commercial time, as well as conduct promotions for business, just as any other commercially-licensed station. It can and will also present local news, sports, weather and do live, on-site remote broadcasting from area businesses for financial remuneration. Community involvement and support of area organizations, community festivals and events shall be a high service, high-profile priority of AM-71.

## Student Evaluation of Participation

All students participating in either station as a laboratory exercise are required to have taken or concurrently be enrolled in the Sales and Underwriting course and be legally qualified to operate the station under the guidelines established by the Federal Communications Commission. Satisfactory student lab participation in the operation of AM-71 shall be evaluated on a weekly or bi-monthly basis by a member of the station management team, or by an individual assigned by them for that purpose. Students participating in the operation of the station shall be provided with written evaluation in terms of the quality and quantity of their work, by the level of professionalism of their conduct, and the level of their creativity and organizational skills. A sample draft of the evaluation form will be included in the appendices of this handbook.

## Potential for Bonuses/Commissions on Underwriting

Although it has yet to be determined, there is a possibility that those students who are enrolled in the Sales and Underwriting course and who prove to be directly responsible for the sale of advertising on AM-71 may be financially remunerated via bonuses or commissions. This shall be determined at such time as the station is on the air and has been operating for some time.

## Remote Broadcasting Talent Fee Policies

In the future, students who participate in presenting live remote broadcasting from on-site locations from a business as part of a promotional event or paid sports event coverage may likewise receive a small stipend to assist them in covering their transportation charges to and from the location of the live remote broadcast. However, until formally changed in the by-laws of the corporation, the only financial remuneration provided to students shall be through bonuses or commissions on the sale of advertising on AM-71 on a token basis.

## Adherence FCC Rules Established in CFR

It shall also be recognized that this station shall, like her FM counterpart, be operated in accordance with all FCC rules and guidelines. This includes the fact that it is currently illegal to advertise any form of alcoholic beverage or tobacco on the radio. In addition, this station, as well as her FM sister station and the Great Lakes School of Broadcasting shall be established as smoke-free, alcohol-free, drug-free environments. Any person who is discovered to be violating these standards shall be summarily dismissed from the Great Lakes School of Broadcasting, Inc. and or the station without due process or entitlement to any portion of refund of their tuition.

## Dedication to Community Service

Through its dedication to community-minded programming, AM-71 shall take this stance a step further and make every effort to support organizations dedicated to the safety and well-being of our community and its citizens such as those which oppose drinking and driving (like Mothers Against Drunk Driving or MADD, Students Against Driving Drunk or SADD, etc.), using drugs (Such as the Just Say No campaign) or smoking as a health hazard (such as the American Cancer Society). AM-71 shall, at all times, project a programming image of positive and good family and high moral values. Whenever such an opportunity arises, AM-71 shall make every effort to present programming which supports the efforts of local, regional, state and national community-service organizations as the U.S. and Big Rapids Jaycees, the Lions Club, the Rotary Club, the Optimists International, the American Red Cross, and all other publicly recognized human service and community service organizations.

## Adherence to School Rules, Policies, and Individuals in Authority Positions

It is likewise important to recognize and understand that, unlike other schools, your work shall be scrutinized by not just your instructors, but by the general public, as well as quite possibly by other potential employing stations who will be in a position to hire you in the future. When you are employed by a station, you shall operate according to their rules. While enrolled in the Great Lakes School of Broadcasting, all students will abide by its rules, and shall likewise abide and conduct yourself in accordance with the rules and programming standards herein established in this policy handbook for AM-71. Any references to or the glorification of the use of drugs, alcohol, pornography, illicit sex acts, or other acts of moral turpitude either directly, indirectly, or through the playing of the lyrics of any song not on the approved stations playlist shall result in immediate dismissal from the laboratory workshop, the station, and expulsion from the Great Lakes School of Broadcasting, Inc. without any recourse of recovery of any portion of tuition or commissions-bonus monies ever owed or earned from AM-71 or Q-92FM.

## Adherence to Laws

Specifically, AM-71 operates in accordance with the regulations established by the Federal Communications Commission, as well as the laws and ordinances of Mecosta Township, Mecosta County, the State of Michigan, as well as all Federally established laws of the United States of America. Any individual who is found to be violating those laws shall be turned over to the appropriate law enforcement agency for appropriate investigation and shall be summarily dismissed from participation in this station as well as expelled without any claim for tuition reimbursement from the Great Lakes School of Broadcasting, Inc. or rights to commissions or bonuses ever earned or owed to them by AM-71 or Q-92FM.

## Theft

It shall likewise be illegal to defraud or embezzle any funds, misrepresent, fraudulently represent, seek to misguide the public or promote any business or advertise an unlicensed lottery on the air, in likeness or in kind in any manner, or unauthorized removal from the premises or stealing of any equipment or materials belonging to the Great Lakes School of Broadcasting, AM-71 or Q-92FM. Penalty for such action shall be the same as the aforementioned: any individual who is found to be violating these rules shall be turned over to the appropriate law enforcement agency for appropriate investigation and shall be summarily dismissed from participation in this station as well as expelled without tuition reimbursement from the Great Lakes School of Broadcasting, Inc.

## Part Two: Orientation to the Stations

### Q-92 FM

Q-92 FM shares an identical mission and statement of purpose to that of her sister station, AM-71. Basically the same aforementioned rules and guidelines that apply to the operation of AM-71 likewise are applicable to those participating in the laboratory workshop on Q-92-FM. The license of Q-92FM is held by a non-profit corporate entity separate from that of AM-71 or the school for legal and accounting purposes. Like her sister stations corporate controller, the daily operation, management and supervision shall be made by two of the primary officers of the controlling corporation, James and Mary Sue McCluskey or, in the event of their deaths, their heirs or assigns. This management arrangement shall hold true unless another individual is so designated by them on a temporary part-time or full-time temporary or permanent basis.

Initially Q-92 FM shall be placed on the air and licensed as a non-commercial, local, low-power FM broadcast service on a frequency of 91.7 MHz. with an initial power level of 250 watts covering an area including Big Rapids, Rogers Heights, Stanwood, and surrounding areas. As an FM operation, the broadcast period for the station is virtually unlimited. However, for convenience, the broadcast service hours of the station shall be determined by the primary officers of the corporation, James and Mary Sue McCluskey. As the station is licensed as a non-commercial, educational entity, it is permitted under existing FCC rules and regulations to operate (as do many school-associated stations) with a flexible broadcast schedule. As commercially licensed full-time stations are required to be on the air a minimum of eight (8) hours per day, excluding daytimers, non-commercial stations operate under different rules and guidelines, which include substantially less than eight hour per day periods. This includes cessation of operation for temporary periods, such as summer vacations and other short-term intervals, such as between Christmas and New Years, Spring Break, etc. Such periods of non-operation shall be designated at the discretion of the station management.

### Non-Commercial Policy

As it is federally licensed as a non-commercial educational (NCE) FM station, it is illegal for Q-92 FM to broadcast any form of commercial advertising. It *MUST* be understood by all personnel participating in the Q-92FM laboratory workshop that this station is completely different from its sister station, AM-71, which is a commercial station but likewise operated by a separate non-profit corporation.

## Underwriting/Program Sponsorship sales Policies

Financial support for the operation of a non-commercial educational NCE station like Q-92FM is derived from the sale of the underwriting of programming, program sponsorship (in whole or in part), from listener support, or from a combination of these. All students participating in either station as a laboratory exercise are required to have taken or concurrently be enrolled in the Sales and Underwriting course and be legally qualified to operate the radio station under the guidelines established by the Federal Communications Commission in the CFR (Code of Federal Regulations). In this course students will learn about the differences in regulations affecting underwriting as compared with advertising. Specifically, underwriting announcements prohibit any form of a Call to Action. This Call to Action would include any suggestion of <Come on out and see us> or <hurry in now for the lowest prices and best selection.>

## Prohibitions in Underwriting/Sponsorship Messages

Past underwriting rules have prohibited mentioning any kind of call to action, prices, specific brand names, or any form of brand name comparison. It is much safer to say the following hour (or whatever time period) or program title is underwritten by a grant from the name of the sponsoring entity and its location. Specifics which are allowed under the ever-changing FCC rules shall be covered in greater detail within the curriculum content of the Sales and Underwriting course.

## Compliance with Fcc Rules and Regulations

It shall also be recognized that this station shall, like AM-71 counterpart, be operated in accordance with all FCC rules and guidelines and all participating in this workshop are legally qualified to do so under current FCC rules and regulations. This includes the fact that it is illegal to advertise any form of alcoholic beverage or tobacco. In addition, this station, as well as her AM sister station and the Great Lakes School of Broadcasting shall be established as smoke-free, alcohol-free, drug-free environments. Any person who is discovered to be violating these standards shall be summarily dismissed from the Great Lakes School of Broadcasting as well as the station without due process.

## Support of Local Community, State,
## Regional and National Service Organizations

Through its programming, Q-92FM shall take this stance a step further and make every effort to support organizations that oppose drinking and driving (such as Mothers Against Drunk Driving or MADD, Students Against Drunk Driving or SADD, etc.), using drugs (Such as the Just Say No campaign) or smoking as a health hazard (such as the American Cancer Society). Q-92FM shall, at all times, project a programming image of positive and good family and upright moral values. Whenever such an opportunity arises, Q-92FM shall make every effort to support such community-service organizations as the Jaycees, the Lions Club, the Rotary Club, the Optimists International, the American Red Cross, and all other nationally and locally recognized human service organizations.

## Scrutiny of Student Work, Attitude and Behavior

It is likewise important to recognize and understand that, unlike other schools, your work shall be scrutinized by not just your teachers, but by the general public, as well as quite possibly by other potential employing stations who will be in a position to hire you in the future. When you are employed by a station, you shall operate according to their rules. While enrolled in the Great Lakes School of Broadcasting, Inc., all students will abide by its rules, and shall likewise abide and conduct yourself in accordance with the rules and programming standards herein established in this policy handbook for AM-71 and Q-92FM.  Any references to or the glorification of the use of drugs, alcohol, pornography, illicit sex acts, or other acts of moral turpitude either directly, indirectly, or through the playing of the lyrics of any song not on the approved station playlist shall result in immediate dismissal from the laboratory workshop, the station, and expulsion from the Great Lakes School of Broadcasting, Inc. without any recourse of recovery of any portion of tuition or commissions-bonus monies ever owed or earned.

### Adherence to Federal, State, County and Township Laws, Rules, Policies and Regulations

Specifically, Q-92FM operates in accordance with the regulations established by the Federal Communications Commission, as well as the laws and ordinances of Mecosta Township, Mecosta County, the State of Michigan, as well as all federally established laws of the United States of America. Any individual who is found to be violating those laws shall be turned over to the appropriate law enforcement agency for appropriate punishment and shall be summarily dismissed from participation in this station as well as expelled without any recourse for any portion of tuition reimbursement from the Great Lakes School of Broadcasting, Inc. and without any recourse of recovery of any portion of commissions-bonus monies ever owed or earned to them by either AM-71 or Q-92FM.

### Theft

It shall likewise be illegal to defraud or embezzle any funds, misrepresent, fraudulently represent, or promote any business or advertise an unlicensed lottery on the air, in likeness or in kind, or unauthorized removal or the stealing of any equipment or materials belonging to the Great Lakes School of Broadcasting, AM-71 or Q-92FM. Penalty for such action shall be the same as the aforementioned: any individual who is found to be violating these rules shall be turned over to the appropriate law enforcement agency for appropriate investigation and shall be summarily dismissed from participation in this station as well as expelled without any recourse of recovery of any portion of tuition reimbursement from the Great Lakes School of Broadcasting, Inc. and without any recourse of recovery of any portion of commissions-bonus monies ever owed or earned to them by either AM-71 or Q-92FM.

### Professional Expectations

All who participate in the operation of both AM-71 and Q-92FM shall be expected to always reflect the highest level of professionalism and integrity. It should go without saying that this is the communication business. Beyond what is said on the air, every time we speak, everything we say, what we do, how we act, how we present ourselves, and how we dress is judged by others.

It is important for the professional credibility of the Great Lakes School of Broadcasting, Inc., AM-71 and Q-92FM that everyone constantly keeps these high ideals first and foremost in their thinking. Other stations may permit ridiculing or teasing of other on-air personalities, but this will not be permitted in these operations.

First and foremost, participants in the laboratory workshops held on AM-71, Q-92FM and students enrolled in The Great Lakes School of Broadcasting, Inc. shall all operate from a position of edifying (or building up) one another. We are all in these operations together. As with may successfully-operated broadcast operations, they are made up of a group of individuals who recognize the importance of a team effort. It is the team which can achieve the goals: individual stars or those who feel they must step on others to achieve success do not belong on this team. When entering these operations, please leave your big heads outside: they are not permitted through the doors of these stations or this school. No Prima Donnas are permitted on the premises. There will always be people greater or lesser than yourself: your personal goal is to do your best, and to help those around you improve and do their best, as well. There is plenty of room in the profession for people who are creative, who recognize and who can do quality work, and who are dedicated to achieving success through the value of teamwork.

The broadcasting business is a cut-throat profession. You may well find that, outside the confines of these operations, many people may well have hidden agendas to demean others to make themselves look better. However, those students who graduate from this program will be prepared to deal with these kinds of individuals and not fall into the trap of thinking of themselves better than anyone else on their team. It is a goal of these stations to help everyone build each other up, to help everyone progress and achieve their own personal potential, and to recognize the value of working in a team atmosphere dedicated to achieving success for the entire operation.

Tattle-tales are not well liked. However, if you observe someone who is violating the rules of this operation or likewise violating laws which could affect the safety, well-being, or integrity of others or of the Great Lakes School of Broadcasting, AM-71 or Q-92FM, it is their obligation to bring it to the immediate attention of their supervisor.

### Specific Professional Expectations
### Student Attitude And Behavior

1. Dress Code. Many stations do not have a dress code. Several successful stations do have a strict and rigid dress code enforced. The Great Lakes School of Broadcasting, Inc., Q-92FM and AM-71 and this school do have a dress code for business attire which you will be expected to follow. While on the premises of either station or the school, professional attire is expected. This is left somewhat up to the judgment of the individual. However, general guidelines would include clean, ironed or pressed slacks and a blouse or a dress for women, and clean, ironed or pressed slacks and a white or colored shirt for men, preferably buttoned down the front. Obviously you are judged by how you look. Men may wish to occasionally add a tie while women may wish to wear a dress appropriate in a business office setting. Professionally appearing socks and shoes should always be worn. We want you to be comfortable, yet conform to professionally-recognized norms for business attire. If you have doubts about whether something may be appropriate or not, it may be best to follow your instincts and not wear it, or bring a change of clothes in your car which you can change into if what you are wearing does not meet with the approval of your peers or station personnel.

Additionally, those who are going off premises to solicit advertising or underwriting should expect to wear suits, especially when meeting area business owners and operators. Suits are available for both men and women. Wearing them will increase your level of success proportionately. Obviously the weather and climate will affect what you wear. Some days a suitcoat or sweater may be appropriate.

Business shirts or blouses are preferred to T-shirts and may look best when tucked in to your slacks or pants. What is frowned upon is generally cheap or holy blue jeans, sloppy shirts, slacks, unclean or smelly clothing, greasy hair, etc. A certain level and degree of etiquette in business attire is expected, even when coming in to do an air shift, for you never know who may drop in to visit the station. What you wear and how you appear is a direct reflection on all of us, and we want our team to appear professional at all times. We don't expect you to buy a whole new wardrobe, but do keep in mind that you represent our stations and our business and how you look is a reflection of all of us. Make an effort to always be clean, neat and well-groomed, always wearing clean and neat appearing clothes that look professional, especially when you are making public appearances as a representative of our stations.

2. Attitude and Behavior. At all times, whether in the stations or outside the stations in public or private, each student shall maintain a positive attitude regarding everything associated with the stations and the community. This shall include going out of his or her way to positively defend the stations as well as their mission, policies, management and other students. Each student must conduct themselves in a positive and professional manner at all times, inside the stations and when in public or private. Each student shall be respectful of their peers and superiors and conduct themselves in a manner which demonstrates and reflects this respect and support of others.

Reflection of Attitude and Respect/Positive Support for Others. No matter what your personal life may be at home, leave your problems outside the station. Don't bring in a bad attitude, but rather always strive to maintain a cheerful, happy demeanor. Be willing to freely give others compliments, praise and, if needed, a positive note blended with constructive, gentle suggestions for improvement. The true leaders will emerge by these kinds of affirming actions.

Elimination of Negative Rumors. Many small stations are rumor mills and can become little Peyton Places. Always be positive and loyal in backing those in charge of the school and the stations. If asked to confirm or deny a rumor, simply say: <that doesn't concern me> or <you'll have to check with them on that, I am only a student here> If you have a question about something or someone, go directly to the source and confront them: don't be a rumor monger, don't try to stir up others or try to stir up trouble. Concentrate your energies on being positive. Always do the best you possibly can and learn all that you can to benefit you for success in our profession.

### Personnel Problem-Solving Procedures

Our policy shall be:

1) If your observation of an issue or another individual's work or actions is positive, please share it with others.

2) If the issue is negative, first go to the individual which the problem involves and make an effort to work out the difficulty with that individual in a positive manner. Approach the individual in a positive manner with a helpful attitude. If these efforts fail, then you should go to their immediate supervisor for assistance and guidance in overcoming this challenge and helping the individual involved.

Unless it is of great personal concern or something that will help or interfere with the smooth operation of the stations, or the efforts of others, the best policy is to keep your thoughts and opinions to yourself. If you keep your nose out of other people's business, you will find that your nose will be a lot less likely to be chopped off quickly. Mind your own business, and do your own job to the absolute best of your ability. Use your time wisely while you are here or wherever you go. Learn all you can, and feel free to frequently ask questions about things that are explained that you still may not understand or procedures and equipment that may be overlooked. We consider these operations to be learning experiences to benefit you, as well as our listeners, and our aim is to help you get as much practical, hands-on experience in as realistic a situation as you can. In this way you will be best prepared for the multitude of challenges that are awaiting you once you walk out of our doors.

3. Telephone etiquette. At any time you may be asked to answer the telephone and, quite possibly, screen the call or take a written message. The following script should be followed:

Good <morning, afternoon, evening> <Great Lakes School of Broadcasting, Inc., Q-92FM, AM-71> this is <your name>. How may I help you (or how may I direct your call).

Many businesses ask the people who answer the phone to screen the calls. This is not to hide from callers, but to be better prepared to serve a caller. Perhaps a file can be pulled so as to more efficiently serve a caller and to appear more professional. It is always best to know who is calling you in advance. It is our policy to always screen all calls and find out who is calling and what organization or business they are representing. If the person they are seeking is out of the building, in a meeting, or otherwise unavailable, ask if you can please take a message. If the caller refuses, at least get a number from them where they can be reached.

If the caller agrees to leave a message, a message pad or plain paper should be near the phone. Legibly write the callers name, business organization, phone number, who the message is for, and their phone number, as well as a brief message on what the call is about. Also be sure to write the time of day and date the call was taken, and your name or initials at the bottom of the message. That way, if there are any questions, someone can clarify who was calling with you. You might also find out when the best time is to call them back.

If, for some reason, the call is of a personal or confidential nature, it is inappropriate to write on the message pad: *Fred Jones from ASCAP called and wondered when you were going to pay his bill.* Screening calls and messages also requires tact and protecting the interests and confidentiality of your supervisors.

At certain times of the year this script may be changed by the station supervisor. This may include: <Merry Christmas or Happy Holidays,>

(Q92-FM, AM-71 or Great Lakes School of Broadcasting, Inc.).

## Appendix A

### *Student Laboratory Workshop Participation Evaluation Form*

STATION:          (circle one)              AM-71                    Q-92FM

Student evaluated:  _ _ _ _ _ _ _ _ _ _ _ _ _ _ _ _ _ _ _ _ _ _ _ _ _ _ _ _ _ _ _ _ _ _ _ _ _

Name of evaluator:  _ _ _ _ _ _ _ _ _ _ _ _ _ _ _ _ _ _ _ _ _ _ _ _ _ _ _ _ _ _ _ _ _ _ _ _

Date:, _ _ _ _ _ _ _ _ _ _ _ _ _ _ _ _ _ _ _ _ _ _ _ _ _ _     199 _ _ _ _ _ _ _ _

#### *Title of Position Evaluated: (circle one):*

DJ-Announcer     Production     Receptionist-Office     Board Operator     News     Sports

Evaluator: (PLEASE CIRCLE ALL APPROPRIATE WORDS BELOW EACH Category Title)

##### *Attitude*

Poor   Negative   Demeaning   Self-centered   Needs improvement   Good   Cheery   Positive

##### *Team-Player*

Poor   Neg.   Demeans others   Big headed   Self-centered   Improving   Good   Affirming Positive

##### *Creativity*

Poor    Negative    Weak    Inconsistent    Needs improvement    Good    Dependable Excellent

##### *Professionalism*

Poor   Negative attitude   Tardy   Self-centered   Needs improving   Good   Has a halo   Excellent

##### *Since Last Evaluation*

N-A     Poor     Negative     Still needs improvement     Improving     Good     Excellent

##### *Overall*

Not likely to pass     Barely passing     Average     Above Average     Excellent     One-of-a-kind

Any Additional Commendations:

Any Additional Recommendations for Improvement:

Overall:  S or U

## Appendix B

*Sales-Underwriting Laboratory Workshop Participation Evaluation Form*

STATION:        (circle one)              AM-71              Q-92FM

Student evaluated:  _____

Name of evaluator:  _____

Date:  _____  199 _____

Evaluator: (PLEASE CIRCLE ALL APPROPRIATE WORDS BELOW EACH Category Title)

*Attitude*

Poor   Negative   Demeaning   Self-centered   Needs improvement   Good   Cheery   Positive

*Team-Player*

Poor   Neg.   Demeans others   Big headed   Self-centered   Improving   Good   Affirming Positive

*Organizational-Creative Planning Ability*

Poor   Negative   Weak   Inconsistent   Needs improvement   Good   Dependable Excellent

*Professionalism*

Poor   Negative attitude   Tardy   Self-centered   Needs improving   Good   Has a halo   Excellent

*Since Last Evaluation*

N-A   Poor   Negative   Still needs improvement   Improving   Good   Excellent

*Overall*

Not likely to pass   Barely passing   Average   Above Average   Excellent   One-of-a-kind

Any Additional Commendations:

Any Additional Recommendations for Improvement:

Amount on books this mo.: $_____   Amount of books to date: $_____

Overall:  S or U

# Chapter 12 ▶ ▶

# Underwriting and Sales

Underwriting is perhaps the most confused, most abused and least used method of funding for noncommercial educational (NCE) radio stations. Yet underwriting can yield the highest potential of dollars for your station, particularly if it is operated by a small school district, a high school, small two-year college, community college or small university.

In the eyes of the Federal Communications Commission and according to its rules and regulations noncommercial means just that. As contrasted with their commercially-licensed counterparts, noncommercial stations are prohibited from broadcasting commercial messages.

## Announcements Promoting the Sale of Goods and Services

The following excerpt comes from the FCC Public Notice 86-161 (36590) dated April 11, 1986:

"Section 399B of the Communications Act of 1934, as amended, and Sections 73.503(d) and 73.621(e) of our rule specifically proscribe the sale of goods and services of for-profit entities in return for consideration paid to the station. These rules, however, permit contributors of funds to the station to receive on-air acknowledgments. The commission has articulated specific guidelines which emphasize the difference between permissible donor and underwriter announcements and commercial advertising. See Commission Policy Concerning the Non-commercial Nature of Educational Broadcasting Stations, 97 FCC 2d 255 (1984) (hereinafter referred to as "1984 Order"); Commission Policy Concerning The Non-commercial Nature of Educational Broadcasting Stations, 90 FCC 2d 895 (1982) (hereinafter referred to as "1982 Order"); Second Report and Order, 86 FCC 2d 141 (1981); First Report and Order and Notice of Proposed Rule making, 69 FCC 2d 200 (1978).

According to FCC Public Notice 86-161.36590: "The Commission has become aware of significant uncertainty and controversy concerning various aspects of Commission and statutory policy relating to commercial underwriting on non commercial stations. As a consequence, we have reviewed the existing policies, focusing on five specific areas: (1) the broadcast of announcements relating to goods and services for which consideration is received by the station; (2) enhanced underwriting and donor announcements; (3) the offering of program-related materials; (4) the practice of hosts of children's programs urging the purchase of program-related materials; and (5) the airing of foreign language programs by public broadcasters and the degree of control necessary to ensure compliance with Commission requirements."

There is a great deal of confusion about what constitutes an underwriting message (which is legal and which can be sold to sponsoring businesses and organizations), and what moves a message from the underwriting side to becoming a commercial announcement or CA. This FCC Public Notice makes mention of noncommercial (NCE) broadcasters having aired outright commercial messages on behalf of profit-making entities in violation of the rules and statute.

## Two Examples of Rule Violations

On October 25, 1985, the Commission issued Notices of Apparent Liability to the licensees of two noncommercial stations for repeatedly airing commercial messages in violation of the Communications Act and the Commission's rules. In addition, a letter of warning was issued that same day to the licensee of a third station for airing underwriting acknowledgments which contained comparative and qualitative descriptions of the donors' products and services and, therefore, exceeded Commission guidelines.

In this Public Notice the Commission affirmed its intent investigate such complaints. It continues: "Information brought to the attention of the Commission regarding such practices will be scrutinized and licensees found to have engaged in them will be sanctioned."

## Enhanced Underwriting and Donor Acknowledgments

In this same Public Notice the Commission acknowledged "that some public broadcasters may be airing donor and underwriter acknowledgments which exceed the Commission's guidelines. . . .In March 1984, we relaxed our noncommercial policy to allow public broadcasters to expand or "enhance" the scope of their donor and underwriter acknowledgments to include: (1) logograms or slogans which identify and *do not* promote, (2) location information, (3) value neutral descriptions of a product line or service, and (4) brand and trade names and product or service listings. *1984 Order* at 263. That action was taken as another step in our ongoing effort to strike a reasonable balance between the financial needs of public broadcast stations and their obligation to provide an essentially noncommercial service. It was our view that 'enhanced underwriting' would offer significant potential benefits to public broadcasting in terms of attracting additional business support and would thereby improve the financial self-sufficiency of the service without threatening its underlying noncommercial nature. In this regard, we emphasized that such announcements would not include qualitative or comparative language and that the *Order* should not be construed as allowing advertisements as defined in Section 399B of the Commission's Act."

## Section 399B

(a) For purpose of this section, the term "advertisement" means any message or other programming materials which is broadcast or otherwise transmitted in exchange for any remuneration which is intended: 1)To promote any service, facility, or product offered by any person who is engaged in such offering for profit; 2) to express the views of any person with respect to any matter of public importance or interest; or 3) to support or oppose any candidate for public office.

In Public Notice 86.161.36590 the Commission reiterated that "acknowledgments should be made for identification purposes only and should not promote the contributor's products, services or company. For example, logos or logograms used by corporations and businesses are permitted so long as they do not contain comparative or qualitative descriptions of the donor's products or services. Similarly, company slogans which contain general product-line descriptions are acceptable if not designed to be promotional in nature." Telephone numbers included in an acknowledgment announcement is within these general guidelines and is permissible.

The Commission also provided several examples of announcements that would violate the rules:

A.    Announcements containing price information are not permissible. This would include any announcement of interest rate information or other indication of savings or value associated with the product. An example is:
- "7.7 interest rate available now."

B. Announcements containing a call to action are not permissible. Examples are:
- "Stop by our showroom to see a model'";
- "Try product X the next time you buy oil."

C. Announcements containing an inducement to buy, sell, rent or lease are not permissible.

Examples of such announcements are:
- "Six months' free service";
- "A bonus available this week";
- "Special gift for the first 50 visitors."

Simply put, an underwriting message must not contain a call to action. This basically means you cannot urge listeners to call, write, stop by, hurry down, or any other form of motivation. Obviously there is a lot of gray area here. However, if you put each piece of copy or message through this close scrutinizing process, you should be able to determine whether or not you have a bonafide underwriting message or a commercial.

In its truest and safest form, programming can be underwritten most clearly by simply stating that the next half hour, next hour, or the following program in total comes to you courtesy of a grant from the XXX organization. There are many variations on this kind of announcement, but the basic gist of it must remain: you must identify the name of the sponsoring organization that is basically paying your station to help bring the program to listeners.

Things become muddled when a business simply cannot understand why your station will not urge shoppers to hurry in because the sale ends this Saturday or why your station has rules about not mentioning products and prices. Your safest path is to limit the information included in an underwriting announcement to the name of the business or organization underwriting the program, their address and a slogan. Again, the slogan must not have any form of a call to action.

An example of how closely you must monitor your messages follows. A message from a business is acceptable, as long as the spot does not become a commercial by urging listeners to take specific commercial action. Something like this program is sponsored by Nike sports shoes is OK, but "buy Nike's new hot pink pumps" would be a commercial, and thus, illegal on a noncommercially licensed radio station.

Station programmers and underwriting managers become even more confused when dealing with nonprofit organizations. Your station may accept money in exchange for time purchased to run an announcement for a bonafide nonprofit corporation, but the message must be logged as a Paid Public Service Announcement. The idea behind this is that you are guaranteeing a specific schedule of announcements will air for a NON PROFIT organization. As a matter of fact, it is perfectly legal to air a full-blown commercial for a not-for-profit organization on noncommercial radio. The important consideration is that you make absolutely sure that the organization is registered as not-for-profit.

Some stations are conservative and air Public Service Announcements (PSAs) only for nonprofit organizations. However, noncommercial stations should not overlook this as a viable option. What follows are some discussions regarding NCE underwriting issues pertinent to this chapter. These are included with permission of each author. It is hoped that this information will be useful, enlightening and informative as you consider the possibility of additional funding to help cover your new station's operating expenses.

*Subj: underwriting #20 10-JUN-1996 21:03:05.23*

Is underwriting considered a tax deductible donation?
Subj: RE: underwriting #21 11-JUN-1996 07:37:44.54
Yes, in MOST cases, underwriting is a tax-deductible donation.

Here are some things to consider:

1.  Is your station, or its owners (e.g. University Board of Trustees) noncommercial by tax definition (501 c3, etc.)?

2.  Are we talking about personal or business donation?
    (a)  If personal, check your state laws concerning any limits, whether financial or to what type of organization.

(b) If business, the business entity may want to check with their accountant(s) to determine whether they have reached the $ limit they can "donate" AND ALSO as to whether it would be better for the business to label it as a tax-deductible donation or as an advertising expenditure. (Don't confuse the limits of your donor announcement with how they list their donation. They can call it what they want, YOU have to follow FCC rules concerning announcements.)

3. Do you have a "Grants" Agreement form for the business to sign which specifies that they are:

(a) Giving the money or gift-in-kind (e.g. CDs from the record shop) as a donation to your station?

(b) Recognizing that this is a donation and that you (the station) will make donor announcements—but are not guaranteeing a specified number of announcements? and

(c) Understanding the limitations of what you can and cannot say or do (stylewise) on the air?

The main reason for this Agreement ("contract", if you will) is not legal but is for up-front clarification of understanding between you and the business as to what this is all about. Over the many years, it is my unavoidable experience that 99% of all businesses do not understand the difference between advertising on the local commercial station and on your noncommercial station and are upset when they don't hear a full-blown 60-second CA at least "x" number of times for the donation they made. Far better to lose a few donors than to have ill will and bad PR running through your territory. Business donors are often tough to get and keep—but well worth the effort if you target appropriately for your listener.—BEST OF LUCK!—Jerry Henderson, Central Michigan University

#2211-JUN-1996 09:11:35.66

It most certainly is. Generally, an underwriting "buy" is just like a charitable donation to a nonprofit organization, which is tax deductible. Your station's on-air announcements merely acknowledge the support. And, as far as I know, advertising on commercial stations is also tax-deductible since it's a business expense. Hope this helps.—Chris Wheatley, Manager, Radio Operations, Ithaca College

#2311-JUN-1996 09:14:32.96

Subj: Underwriting -Reply
Let me be the first of probably many to say that the answer is "yes"—Allen Myers, amyers@fcc.gov, (202) 418-2774

*#2411-JUN-'96 11:10:10.91*

Subj: Underwriting: churches
    As long as we are discussing underwriting, are there any specific
rules regarding underwriting paid for by religious institutions? We have
a nondenominational black gospel show that wants to begin soliciting
contributions, and it looks as though it will be very successful.—Brian
Gawor, G.M., WVKC 90.7 FM, Knox College, Galesburg "There is
absolutely no inevitability as long as there is a willingness to contemplate what is happening."—Marshall McLuhan bgawor@knox.edu


*Subj: Underwriting: churches -Reply*

This message warrants some discussion. As long as the sponsoring
organization is a bonafide nonprofit organization, Brian can air its
solicitations like he would for any other charitable organization—provided that the solicitations do not interrupt regular programming.
[From Brian's post, it appears that the solicitations will be made in the
course of the show, so this may not be a problem.] If there is to be a
problem, it will come in Brian's determination which programs to air
with this type of announcement. If he tries to permit some organizations to do this and not others, he might receive objections from the
organizations which he has denied. Therefore, my recommendation
would be that he establish a clear station policy on this issue that organizations must adhere to before he accepts their programming.—Allen
Myers, amyers@fcc.gov.


*Subj: Re: Underwriting #205 12-NOV-1996 17:31:05.90*

Okay, as we get detailed with underwriting, can the following
terms be used? 'variety' . . . 'much more' or are they too qualitative?
Additionally, can slogans be used, even if they imply some quality?

Subject: Re: Underwriting
Date: Tue, 12 Nov 1996 From: jmunson
<jmunson@STAFF.UWSUPER.EDU
    The word 'variety' itself is not qualitative, but it would depend on
the context. Something like "Support for KXXX is provided by Mom &
Dad's Pretty Good Hardware, offering a variety of power tools for professional use," is probably O.K. Saying something like "has the biggest
variety" would be against the rules. Saying: 'much more' Red
Alert...Shields up Mr. Worf. Slogans can be used, even if they are qualitative, IF it can be shown that these slogans are a long-time part of the
corporate identity. "GE . . . we make good things for life."
"Smuckers...with a name like Smuckers . . . it has to be good." These are
examples of slogans which, though qualitative per se, are legal because
they are an integral part of that corporation's public identity.
    —John Munson, KUWS Radio

*#210 13-NOV-1996 05:52:57.75*

Subject: Re: Underwriting
. . . and for the most part, they are copyrighted slogans, and there-fore can be used. The "safe" option is always to skip the slogans!!

*#212 13-NOV-1996 08:20:56.18*

From: Warren Kozireski <WKOZIRES@ACSPR1.ACS.BROCK-PORT.EDU>

Subject: Re: NACB FS: Re: Underwriting
At the NACB National just concluded, Attorney Carey Tepper again handed out a booklet with common do*s* and don't*s*. The logo phrase example he used as being okay was GE—we bring good things to life. The example used as being illegal was "Get Met-It Pays". I'm sure the NACB office can get you copy of the handout if you call.

## Summary

The Federal Communications Commission is very serious about enforcing the rules which prohibit commercials from being aired on a noncommercial educational NCE station. Commercial stations in your area can easily be threatened by a new NCE radio station that decides to extract extra dollars from their community through underwriting.

## Conclusion

Make absolutely sure that everyone understands the rules for properly structuring a donation acknowledgment message. Also have everyone understand the elements that turn an underwriting message into a commercial. You should: post the "rules" and guidelines for underwriting announcement content in a prominent location in your new station. And good luck.

# Chapter 13 ▶ ▶

## MOTIVATING YOUR STAFF FOR QUALITY

---

Marshall Field said "Goodwill is one asset that the competition cannot undersell or destroy." People are motivated to do things for a great variety of reasons. Typically in business it is profit, whether that be personal financial gain or corporate profit. In student radio, there is seldom personal financial gain to be achieved by a staff member. Many student stations operate with all or nearly all volunteer staff members.

Most students volunteer to work at the school radio station for two primary reasons: either they want to learn more about radio or to have fun, or a combination of both of these reasons. These volunteers certainly are not helping because there is money involved, but there is personal gain. Student radio station workers earn a sense of satisfaction and achievement in knowing they are all working together to produce a quality product that is enjoyed by many people. These volunteer workers also realize that the skills they are learning will be directly transferable to a job in the industry if they continue to work hard and learn.

Nearly all of these workers will need guidance. This guidance can come from paid or unpaid staff supervisory personnel. The delegation of authority is determined by how you have organized your management structure. Usually station policy will dictate the rules for station operation and personnel behavior. Typically the majority of the staff members will respect and obey the station rules.

The types of students working at your new station will probably fall into one of three categories. First, you will have the valuable, hard-working volunteers who spend a lot of time working at the station to improve it. They take an active interest in station operations, programming, and seeing to it that everything is going smoothly.

The second type of student is only in it for themselves. They volunteered to help out because they just want to have fun and earn their share of fame and fortune. There is nothing wrong with this, as long as these students continue to follow the rules and policies of the station.

The third type of student volunteer might be classified as the kind of student who rebels against authority of any kind. These students like to push the envelope and find out just how much they can get away with before they are caught. It is not always easy to determine which category a volunteer may fall into, but sooner or later you will make this determination.

## Common Motivational Problems

Often students have a multitude of activities and responsibilities outside the station. For some, first and foremost is their education and their studies. Homework continues to be assigned, papers are required, and tests must be taken and successfully passed by students. The more active students will also be involved in other extracurricular activities in addition to working at the radio station. These activities may include athletic competition, the arts, sciences, as well as family and personal relationships. Some of your staff members may be on the school football, basketball, baseball, track, soccer, gymnastics, volleyball or swim teams. Other students may participate in marching band, orchestra, or choir. Some may also belong to the French or Spanish club while others are required to attend meetings and participate in such after school activities as the National Honor Society, student government, student newspaper, or a multitude of other group responsibilities.

Time is valuable to any active student. Some of your station volunteers may simply not have the time that they would like to dedicate to your station. Yet they still want to be involved, at least on a minimal level. This may include doing a shift or two per week.

It is a unique challenge to try to motivate these kinds of people into doing quality work while they are under severe time limitations.

Quite often it is a matter of practice and training. The more these kinds of students do production, the more skilled they become at the task. By offering them plenty of guidance, support and help as they need it, you will find that they will develop into first class workers.

It is critical that you explain and demonstrate what is expected of them, and what you consider to be quality work. As they observe and learn for themselves what goes into producing a quality announcement or doing a quality air shift, for example, they will strive to produce a higher quality product for your station.

## Facing the Challenge

The problem occurs when you have a staff member or group of workers who will not follow the rules and guidelines established by the station advisor(s), manager(s) or department head(s). It is difficult to admonish or punish a volunteer when you are not paying them for their efforts. This is one advantage commercial station operators have over their noncommercial counterparts.

However, you must remember what factors are motivating these people to participate in your station operations. It must be made clear to all volunteers from the outset that working at the station is a privilege, like driving a vehicle. If the privilege is abused, then it can be suspended or revoked. It must be the same way with their privileges of working at the station.

If they are there to learn and have fun, then working at the station is important to them. Having the privilege of working at the station taken away for a short period should have a profoundly positive effect on any volunteer who is sincere about wanting to remain associated with your station.

Finally, if a volunteer really does not care, then suspending or permanently revoking their privilege of working with the station really does not matter. In these unfortunate situations, it is usually best to take the extreme measure and politely invite them to resign their position or duties. Having negativism or pessimism in your ranks can prove to be disastrous, and it is usually best to remove these people from within your organization. It is also important that you take measures to see that they stay out, and are not in a position to do harm to your operation.

## Motivational Factors

People are motivated to do things for a great variety of reasons. Sometimes it is the fear of failure that drives them forward. Rejection and disappointment in themselves or from others can also have an effect on the human psyche.

Success, reward and admiration from one's peers can also be a motivational force. When asked what motivates her, a high school senior replied: "I think about how good its going to turn out at the end. I think about the reward." While reading a book by Dr. William Glasser titled *Positive Addiction,* another "noted psychologist" came to mind: Dennis E. Waitley, Ph.D. While the author and his wife owned the radio stations, we often found ourselves challenged to motivate our staff, particularly our sales staff. To assist us in our motivational goals, we purchased Waitley's cassette tapes on motivation: "The Psychology of Winning: Ten qualities of a total winner." In *The Psychology of Winning,* Dennis E. Waitley's ten basic principles or "Qualities of a Total Winner are: 1) Positive Self-Expectancy 2) Positive Self-Motivation 3) Positive Self-Image 4) Positive Self-Direction 5) Positive Self-Control 6) Positive Self-Discipline 7) Positive Self-Esteem 8) Positive Self-Dimension 9) Positive Self-Awareness, and 10) Positive Self-Projection.

Much of Waitley's presentation dealt with Positive Self-Expectancy (an overall attitude of personal optimism and enthusiasm) and, particularly crucial in my opinion, Positive Self-Motivation (winners dwell on their desires, not their limitations, and people will change happily and effectively when they want to). We were impressed with Waitley's motivational tapes: at least they always had an uplifting, positive motivational effect on me. Since I personally increased my sales substantially by listening to the tapes, we always believed they were a worthwhile investment.

In *The Greatest Salesman in the World,* Og Mandino summarizes this well on page 26: ". . .he who has never failed is he who has never tried." One needs to be motivated in order to try, and typically this motivation is internally generated. The best incentive is personal: the desire to succeed, to achieve recognition, to get rich, or to aspire toward a deep personal goal or conviction.

A great similarity between Waitley's *Psychology of Winning* and Glasser's *Positive Addiction* is their numerous examples of sports figures (and their associates (Glasser's World Series of 1974 where the Oakland A's easily beat the L.A. Dodgers, and Waitley's (World Series: New York Yankees/Milwaukee Braves; Warren Span, the great Milwaukee left-handed pitcher . . . why would anyone motivate someone else with the reverse of an idea?). Waitley also motivates numerous other sports greats including Lee Trevino!

In *Positive Addiction* (pg. 8) Bill Glasser makes the point "Most people who give up tend to stay away from people who succeed." This author finds the statement easy to agree with based upon his personal experiences in having taught adults on Michigan's welfare roles for two years.

Many of these people, who at first simply seemed shy, tended to be easily intimidated by their classmates who were stronger achievers. Many of these same people were also afraid of success and, as Glasser continues on page 9: ". . .they don't believe they have the strength to succeed. Furthermore, because they are weak they tend to blot out of their minds what they might do to get stronger; they settle for a minimal life because they do not have the strength for a better one." This is a very true statement.

Many people have already put their lives on the track of Bill Glasser's *Positive Self Addiction.* This author has always been able to find an inner strength, whether it is derived from personal, deep religious beliefs, or an inner self-confidence. Many successful broadcasters are successful owners, managers, and sales people because they believe not only in their product(s), but primarily in themselves and their own abilities. As Og Mandino points out in *The Greatest Salesman in the World* (pages 26, 45 and 49): "Failure will never overtake you if your determination to succeed is strong enough."

Mandino's book offers a great many points to be lived for both personal and professional improvement and fulfillment. Among these points is: "...I will seek constantly to improve my manners and graces, for they are the sugar to which all are attracted (page 71)." This point, as are many of his others, could well be taken and used by all for personal as well as professional improvement. This statement brings back memories of the attributes of a friend who had been an instructor for the Dale Carnegie Institute. This friend would most likely agree that the previous quotation is an important key toward self motivation and improvement.

As a broadcast station manager or owner, you will be motivated by Mandino's statements such as this one on page 71: "I will concentrate my energy on the challenge of the moment and my actions will help me forget all else. The problems of my home will be left in my home." How many times our employees (and, yes, even students!) have been told to leave their personal problems at home. How many times has the author's father said "If at first you don't succeed, try, try again?" (Mandino, page 66)

It is strongly suggested that you read and absorb many of these points Mandino makes in the scrolls, and employ them in your own life. Mandino's book will "bring them all together" for you.

Many have asked, "Why do you kid around so much?" This author has always enjoyed humor, laughing and having fun, but has really never fully understood why. It has always seemed a mystery until after reading chapter fourteen of Mandino's book: "And so long as I can laugh never will I be poor (page 86)." Mandino makes reference to laughter being as the life of a child, and only through the eyes of a child will we enter the kingdom.

Chapter fifteen (dealing with multiplying your worth a hundred fold) strongly emphasizes the importance of goal-setting. Through his own life and career, this author has always established realistic, achievable goals that challenge his abilities.

Likewise, this author has always taught the importance (and practiced it) of raising your goals (or setting new goals) once you have met them. To set new goals is to continue to live! As Mandino wrote: "I shall not live in the past! and "I will act now" are important words for everyone to live by: words many have taken all too seriously and accomplished tasks to the surprise or envy of their peers. The old cliché: "Don't put off until tomorrow what you can do today" really rings true. Shun procrastination, and urge others on to accomplish their goals.

Many broadcast station managers and owners perform better under stressful or challenging conditions. If there is no deadline or self-imposed goal to accomplish a task, some people tend to procrastinate. Victor Frankel, noted Psychiatrist who wrote *Man's Search for Meaning* flatly states: "What a person really needs in life is not a tension-free state, but the striving and struggling for a goal that is worthy of him or her."

An impressive gentleman named George Logan, General Manager of WIBW-TV Channel 13 in Topeka, Kansas was mentioned in the previous chapter. When he visited our Electronic Media Management class at Kansas State University, students asked him about how he deals with people who complain. His answer was very similar to the positive manner we anticipated him to use in his "walking around/open door" style of management technique. Basically, he discussed the matter with the individual and "motivated" (helped) them to find another position in a larger market (where they could do more complaining, without affecting his operation).

It is very difficult to build or even maintain staff morale if there is a staff member who is a groaner or complainer. In his book titled Positive Addiction (pp. 80–82) Dr. William Glasser explains this situation: ". . . since it is a characteristic of the weak to criticize others or complain a lot, both of which quickly end up as self-criticism, the weaker they are the less they experience the PA (Positive Addiction) state. Without it they have fewer options,. . . so they use these same options over and over. Since happiness ordinarily demands much more strength than these overused and inadequate pathways can provide, the weak are less successful in their quest for love and worth and they hurt."

It is understandable that those who are "weak" on your staff would attempt to point out the mistakes and shortcomings of others, if for no other reason that to draw the attention away from their own weaknesses. Dr. Glasser continued: "Weakness, which is almost always associated with pain, is the exact opposite of strength. The weak person has few options or has little access to the ones he has, which amounts to the same thing. When a person learns through experience that he cannot find the love or worth that he would like to have, the pain tells him to get going and try harder, to consider a new approach, do something. Unfortunately, because he is weak, his options are limited and he quickly runs out of new approaches. He usually ends up trying one or two ways to get what he wants over and over again as if no other choices were possible.

You see this in weak children trying to amuse themselves. They try once or twice, give up, and then get upset and begin to cry, which is their symptomatic way of asking for help. Later in life they may complain that their marriage is unhappy, job no good, house inadequate, the weather bad, their family unappreciative, complaining continually but doing little else." If you have (or had) the unfortunate experience of having one of these habitual complainers on your staff, this author can certainly sympathize with your plight. He has had his share of these individuals and they are difficult to deal with in any situation, particularly when on your staff.

It would appear Dr. Glasser has experienced them, as well, as he explains on page 82 of Positive Addiction: "The constant griping and complaining that is characteristic of the weak drives the people around them crazy. As this happens, the stronger withdraw, leaving the weak person with less love and less chance for someone to believe s/he is worthwhile, effectively compounding his inadequacy. Nothing is more characteristic of a patient seeking psychiatric help than to complain bitterly and say: 'I don't know what to do. I can't seem to do anything, doctor. Help me, I'm so miserable.' Such people feel helplessness and pain because, in contrast to the strong, it is each neuron for itself within their brains. Theirs is a mental house divided against itself. Few of their neurons work well together and what hookups they have are either overused or, because of their self-criticism, unavailable to them."

However, Dr. Glasser explains that this choice works only for a while. "They clearly understand the cause of the pain; that is why they give up. The active pain of giving up may be exactly the same as the pain of a second-choice symptom. For example, we may choose to depress when we give up and we may suffer later from the symptom of depression." Dr. Glasser continues: "It is the same kind of pain but the difference is that when it is a symptom we less and less relate the pain to what we gave up . . . *how it hurts, how it causes pain, is because it is essentially a one-track-brain experience.* It is an attempt to use one small set of pathways over and over until these literally begin to hurt from overuse." Dr. Glasser's advice is to treat the individual ". . . as non-critically as possible so that s/he can accept her/himself and gain more access to her/his mind. We may not ever be able to be her/his best friend or provide anywhere near the love they need—that would be unrealistic. We can, however, be enough of a non-critical friend to help them begin to figure out their options to solve their problems. They have the mental wherewithal, we have to help them to use it and get more."

As you can infer from this passage, it takes a great deal of time, patience and understanding to "reach" this kind of individual. Often, since broadcast station managers are already "overstressed" in dealing with their own myriad of responsibilities, they don't have the time or patience to deal with the psychological problems which sometimes face specific staff members. Thus, it's usually more efficient to "dump the problem", rid themselves of the complainer, or transfer the problem to someone else, which is completely understandable.

It is a shame that the "complainer" often "can't see the forest for the trees" and doesn't realize they have the problem until they've alienated everyone around them, including their co-workers and even their family and loved ones. Dr. Glasser continues by describing a third choice or state these individuals may choose, which is to become "negatively addicted" which is quite the opposite of a positive addiction experience. Please don't infer that becoming a successful broadcast station manager requires a degree in psychology. However, it is beneficial to understand why people act in certain manners, and sometimes apply a little psychology to reach them. This can be most beneficial in dealing with personnel challenges from your staff, or in dealing with clients, disgruntled audience members, other media representatives and other people outside your station.

To relate this knowledge to your own experience you may refer to Og Mandino's book *The Greatest Salesman in the World,* (page 61): "...in silence and to myself I will address him and say I love you." As Jesus Christ said, "You shall love your neighbor as yourself." This is the way you should live your life and manage others in your charge.

Although it may be simplistic and have obvious religious foundations, this is still a very important management principle to remember, practice, and to live by: to love your fellow brothers and sisters, treat them with respect, take into account their needs and adjust your actions and behaviors to accommodate them. These are "your people," and the successful station owner or manager will do everything in his or her power to make their lives easier and more meaningful. Like the caring coach, a broadcast station manager or owner should always try to help their players aspire to achieve their potential. Just like a championship football, baseball or basketball team (or in any team sport), there should be no specific "star": everyone works together for the common good and to achieve the common goal.

Just as any chain is only as strong as it's weakest link, any team is only as strong as the links that bind them. Everyone on your team must have a clear understanding of your station's goals. In addition, they must also have a crystal-clear understanding of their role in achieving the goals. As W. Edwards Deming's theories were so clearly demonstrated to the American automobile manufacturing industry by Japanese auto manufacturers, workers who fully understand their role and their part in the overall plan are much more effective on the job.

Everyone on your team, from the janitor to the General Manager and owner, must completely understand your corporate goals and their own role played in achieving them. As Dennis Waitley states in "The Psychology of Winning" and Quality #9: "Winners know who they are, what they believe, the role in life they are presently filling, their great personal potential—and the future roles and goals which will mark fulfillment of that potential."

In fact, Waitley's portrayal of winners is crucial to this discussion. In the first chapter titled "Positive Self-Expectancy, Quality #1," Waitley states: "The most readily identifiable quality of a total winner is an overall attitude of personal optimism and enthusiasm. Winners understand the psychosomatic relationship: psyche and soma: mind and body—that the body expresses that the mind in concerned with. They know that life is a self-fulfilling prophecy; that a person usually gets what he or she actively expects. Your fears and worries turn into anxiety which is distressful . . . the production of hormones and antibodies changes; resistance levels are lowered and you become more vulnerable to disease and accident . . . if your mental expectancy is healthy and creative, your body will seek to display this general feeling with better health, energy and a condition of well-being."

In Waitley's second quality, "Positive Self-Motivation," which is the key to this chapter in Broadcast station management and ownership, he states that "Winners dwell on their desires (rewards of success), not their limitations (penalties of failure). . . . Fear and desire are among the greatest motivators. Fear is destructive, while desire leads to achievement, success and happiness. (Winners) . . . focus their thinking on the rewards of success and actively tune-out fears of failure. Individuals are motivated by their fears, inhibitions, compulsions, and attractions. We always move in the directions of our currently dominant thoughts. People will change happily and effectively when they *want* to."

In Waitley's third quality, "Positive Self-Image," he points out that winners are always aware of the importance of their self-image." His fourth quality, "Positive Self-Direction" emphasizes the importance of clearly defined, constantly referred to, game plans and purposes. They know where they are going every day, every month, every year. Their objectives range all the way from lifetime goals to daily priorities. Clearly defined, written goals are the tools which make purpose achievable. The reason most people never reach their goals is that they don't define them, learn about them or even seriously consider them as believable or achievable: they never set them, and thus, fail by default."

Waitley's tenth quality, Positive Self-Projection states: "Winners specialize in truly effective communication, taking one hundred percent of the responsibility not only for sending information or telling, but also for receiving information or listening for the real meaning from every person they contact. Winners are aware that first impressions are powerful, and that interpersonal relationships can be won or lost in about the first four minutes of conversation."

In Quality #8, "Positive Self-Dimension", Waitley states winners "have learned to know themselves intimately. They have learned to see themselves through the eyes of others . . . and they have learned to be aware of time—their opportunity to learn from the past, plan for the future, and live as fully as possible in the present. Winners plant shade trees under which they know they will never sit."

It is obvious from this stimulating discussion of the qualities of a winner that simply thinking like a winner is not enough. To be a winner, you have to know yourself and expect to be a winner. This can be extended to your entire management team and station staff. Your job is primarily motivational: the General Manager who sits back and "super-vises" like WKRP's Arthur Carlson is just kidding himself/herself. Being a broadcast station manager or owner is a full time-an-a-half over-time job.

You are the driving force that draws your staff and your station to constant success. You are the flame: the motivator that lights everyone's spirit to achieve excellence. You expect your station to win, to make big profits: you must expect nothing else.

The winning begins with your attitude about yourself and your potential. Thinking like a winner is not enough, you have to expect to win and succeed, both individually and collectively.

The future manager will capitalize on these motivational qualities and further instill them in her/himself and their staff members, both individually and collectively. The future manager will expect to win and be successful. This individual will learn that, according to Waitley, "per-haps more than any other quality, positive self-esteem is the door to healthy high achievement and happiness."

The manager of the future will understand and practice the theo-ries of W. Edwards Deming, motivating each individual by communi-cating a clear understanding of the individual's role in the overall suc-cess of the operation.

You would be wise to study and acquire Waitley's Ten Qualities in the *Psychology of Winning* (1978). The complete program is available through the Nightingale-Conant Corporation, 7300 North Lehigh Avenue, Chicago, Illinois 60648, or by calling (312) 647-0300 or 800-323-5552.

What you begin to learn, feel and think today will influence you for the rest of your life. As Waitley states, you must *WANT* to change, to learn, and to succeed. You must *WANT* yourself and your staff to be winners and expect nothing less. Waitley believes "it's not so much 'what happens' that counts in life; it's 'how you take it' that counts. Each of us has many more choices and alternatives than we are willing to consider."

Pages 96–97 of *The World's Greatest Salesman* suggest: "Only action determines (your) value in the market place and to multiply (your) value, you must multiply your actions . . . walk where the failure fears to walk, work when the failure seeks rest, talk when the failure remains silent, and act now. Hunger for success, thirst for happiness and peace of mind; command and obey your own command. Success will not wait. If (you) delay she will become betrothed to another and lost to (you) forever. This is the time. This is the place. (You) are the person . . . *ACT NOW!*"

"Self-discipline alone can effect a permanent change in your self-image and in you. Self discipline is the winning edge that achieves goals: the mental practice—the commitment to memory of those thoughts and emotions that will over-ride current information stored in the subconscious memory bank. And through relentless repetition, the penetration of these new inputs into our (minds) will result in the creation of a new self image. Remember, losers say: 'How can you expect me to do it? I don't know how!', while winners boast 'Of course I can do it! I've practiced it mentally a thousand times.'" Winners really do make it happen for themselves. Winners are strong self-motivators, and *YOU* are a winner! Start expecting to win NOW! Believe in others as you believe in yourself and you will be motivating successful goal-oriented quality winners.

# Chapter 14 ▶ ▶

## Free Resources

### People/Organizations Formed and Available to Help You

There is a multitude of people and organizations formed to offer low cost or no-cost assistance in your quest to start a new student radio station. The real key is determining who they are and how to contact them. Beyond that point it is a matter of accurately formulating and communicating your question(s) to them, receiving their response, and determining if their answer(s) are accurate and applicable to your project's needs.

One nice thing about people who are involved in student media is that they are usually approachable and willing to provide free advice. You are certainly not the first to start a student or NCE station. For each of the 1800-plus NCE radio stations on the air, everyone had to go through what you are doing. In fact, there are certainly a lot more than that number if you consider stations that are unlicensed carrier current, unlicensed Part 15 operations, leaky cable, cable FM and PA system stations.

Each station went through the same process and growing pains, more or less, that you are now experiencing. Each and every one of these stations is unique. There are similarities and differences across the board in terms of who holds the license, who controls the station, its management structure, programming, mission/purpose, goals and objectives, financial support, facilities, equipment, and a host of other variables.

Even though each station is unique, this is to be considered a very positive attribute. The differences in each station reflect the intent of the operators to serve the mission of the station, whether it be to entertain an audience, to provide alternative music and programming, to train students and individuals who are interested in entering the broadcasting profession, or to serve as a laboratory associated with an educational program in a school, college or university. The diversity of all of these stations is their strength. Though they are diverse, there are enough common elements and similarities among these stations that they can all serve one another and help fill the needs of each other.

When you network with the people involved in these operations and "tap in" to the right communication channels, you will usually receive a plethora of answers to each question that you ask. Often the answers will be similar (particularly if your question deals with a legal or FCC rule making issue). Sometimes you will receive different answers to the same question as a result of the different backgrounds and experiences of the individuals who are answering your question.

In some cases it is necessary to join an organization before many of these free or low-cost resources can be available to you. In other situations you might have people right in your own back yard who have professional or noncommercial radio experience available which you will not be aware of without asking. Asking for help is the key. You do not know and do not learn unless you ask for help.

The key to resource acquisition is networking. Getting to know people. This includes getting to know the people in your school, college or university. Communicating your needs is a primary element in networking. Use the grass roots method of word-of-mouth. Take advantage of the media that are already established in your school such as the school newspaper. Advertise in the paper for volunteers.

It is also important that you network outside your school, within your community and state. It is very likely that your state already has established some kind of state broadcasting association. By joining and getting involved, many doors will open to resource people who can help your fledgling operation get off the ground.

There is also a large number of national organizations which will prove to be valuable resource guides as you accept your challenge. Many of these are included in the appendix of this book.

Usually networking can lead you to a friend-of-a-friend, or someone's father or mother, aunt or uncle or grandparent who was formerly involved in the radio industry. No matter what their experience, it is important that you be polite and listen to information and assistance they have to offer to you and your station members. It is particularly important that you involve your school's administration, as well as teachers, professors, parents and community leaders in the planning and development of your project.

Starting a new student radio station is a major task and you will find you can use nearly all of the help that you get. Keeping a student radio station on the air with a minimal number of problems can sometimes prove to be a major challenge, and that is where a large number of resource personnel at your ready disposal can prove indispensable.

Try to avoid refusing help unless you simply believe that accepting it would prove to be counterproductive to your efforts. Learn to delegate your authority and to manage your time and resources efficiently.

## Getting Caught in the Web

Some of your staff members will most likely be very computer literate. The computer as a tool has proven to be a very efficient method of accessing information, databases and ideas. There are several very good resources available to you and your station personnel if you or your institution has access to the internet. Several Lists are alive with a multitude of questions and answers being transmitted and exchanged every hour. Some of these may pertain to your station. In other cases, you may wish to ask your own question and then wait to receive a variety of different views and thoughts addressing your inquiry. Some sample questions that are commonly asked by people who are interested in starting student or NCE stations are included in Chapter Fifteen of this book.

Three organizations are predominant throughout the country and are specifically developed to help student media and student/NCE radio stations. If your station is going to be associated with a school, college or university and is in any way linked to broadcast education, you should seriously consider joining the Broadcast Education Association or BEA. The BEA will prove to be an invaluable asset to you as you get involved in student radio broadcasting and broadcast education.

## The Broadcast Education Association or BEA

The BEA is organized to assist broadcast educators in providing support and an exchange of ideas, philosophies, theories and assistance. The BEA helps students in these same ways. It also provides numerous scholarships and awards in areas such as production and research. Although this is not a "free resource," the resources are without cost to individuals and stations which are members of the organization. The annual membership fees are structured as represented in the table which follows.

BEA Annual Membership Fees*

Individual$70

Undergraduate**/Graduate Student**/Emeriti$30

*as of the time of this writing

**copy of current student ID is *required*

Another advantage to membership in the BEA is that of receiving reduced registration rates for its annual convention held in the Las Vegas Convention Center each year in April. This convention is scheduled immediately prior to the granddaddy of them all, the Annual National Association of Broadcasters (or NAB) convention. According to some reports, the annual NAB convention is the largest annual gathering that is held in Las Vegas. The BEA convention typically begins on the first Thursday in April and runs through the following Sunday. The NAB convention usually runs through the following week. The national convention rates for BEA in Las Vegas are represented in the table which follows.

BEA National Convention Pre-Registration Fees*

| | |
|---|---|
| Academic Member | $90 |
| Non-Members | $200 |
| Student Graduate**/Undergraduate** | $60 |
| Student Non-Member Graduate**/Undergraduate** | $100 |

*as of the time of this writing
**copy of current student ID is *required*

Another good reason for attending the BEA convention is the pass to the NAB exhibitors hall and some of the NAB sessions that are related to broadcast education. This benefits students and educators alike in providing them the opportunity to see all of the latest broadcast equipment demonstrated during every hour of every day that the convention is open. Your access to "free resources" and the networking possibilities are only limited by how long your feet can hold out, how much energy you have, how outgoing you are, and how much you want to meet others who are doing the same thing that you are doing or that you want to do.

This annual trek to Las Vegas in the Spring is a veritable "who's who" of broadcast educators, broadcast programs, and students who are going to be the "movers and shakers" in the industry. If there is only one broadcast convention that you can attend each year, the BEA convention should be at the top of your list.

At such conventions you can get your questions answered directly in a face-to-face, personal manner by the people who have done it successfully. Many who attend the BEA convention have first-hand experience in broadcast education, student laboratories, and/or advising student media (both radio and television).

Among the other valuable assets you receive from joining the BEA is its an annual membership directory which lists every person who is involved with the organization as a member and participant. If you attend the conference you will receive the BEA convention program, as well as a program of those education-related sessions that are organized and presented as part of the NAB conference. These programs list all of the many sessions which are designed to educate, enlighten and address many of the issues and concerns of people who are involved in the broadcast education industry. Sessions are presented in a variety of formats intended to match the nature of the material that is presented. Typically session formats include panel presentations, research papers presented by individual or multiple authors, poster sessions of noteworthy papers, and roundtable discussions. If you are a broadcast educator who is interested in presenting research on a topic related to broadcast education at the national convention, you want to consider various "calls for papers" throughout the spring and summer for the following year's convention. This author has made every effort to present research in broadcast education and advising student stations at the BEA conventions. A recent paper dealt with how the Internet serves faculty advisors of student radio and television stations. Specifically it was a content analysis of messages exchanged on a popular ListServ which was arranged by topical area.

Membership in such an organization is clearly valuable. You should definitely investigate the advantages and opportunities that membership in the BEA affords you. You can contact the BEA via their web site at http://www.beaweb.org or e-mail them at: fweaver@nab.org, Voice: 202.429.5354 or FAX: 202.775.2981. The BEA's address is: 1771 N Street, NW, Washington, D.C. 20036

Many states have their own organizations resembling the BEA. Joining this kind of "regional" group puts you in touch with other member schools within your own state. Some of these groups hold regular meetings, elect officers, and may even hold competitions for student projects. You would be well advised to find out whether your state has its own BEA group. This information can be determined by calling the BEA. Contact information for the BEA in terms of its address, telephone voice number, fax number, and electronic mail address is contained in the appendices as well as elsewhere in this book. Contact information for other organizations which can serve as resources is likewise located in this book.

In addition to the BEA, there are other noteworthy organizations which you should consider joining. These organizations are not focused on any one particular aspect of student broadcasting and are best suited to handling a wide variety of inquiries.

The two "biggies" that have long served student radio and television stations are the National Association of College Broadcasters, or NACB, and The National Broadcast Society—Alpha Epsilon Rho, also known as "AE Rho." Although they both serve student media, each organization is quite unique and really cannot easily be compared to the other.

## The National Broadcast Society—Alpha Epsilon Rho (NBS/A.E. Rho)

AE Rho is by far the older of the two organizations. Many broadcast faculty members see it as a broadcasting fraternity, with schools forming "chapters," not unlike the Greek communities on campus. AE Rho invites both individuals and schools to join its ranks. Although membership in the number of chapters have declined in recent years, AE Rho remains a vibrant and very viable organization dedicated and available to assist and support your broadcast education efforts.

# The National Association of College Broadcasters or NACB

The other premier student media organization in the United States (and world wide) is the National Association of College Broadcasters, or NACB. Headquartered in Providence, Rhode Island, the NACB offers a multitude of services to its members. Unlike AE Rho, NACB membership is organized by individuals and/or stations, rather than chapters.

The National Association of College Broadcasters was started by four members of the Management Board of BTV, Brown University's student television station, in the spring of 1988 from a three year grant given by the CBS Foundation. Three of the Board members went on to run the association full-time after they graduated. For many years their office was located in Brown University's Alumni relations building at a reduced rate, but this has just recently changed.

NACB was founded to help encourage college students to develop their creative efforts at college radio and television stations and to provide an outlet for the exchange of ideas and information. America's 1,400 college radio and 800 college television stations largely functioned in isolation before NACB. There were no comprehensive trade publications oriented toward the college budget and expertise.

Today, the NACB is divided in to five major areas, including: Conferences, Finances, Marketing, Memberships, and Networks and programming. You will find valuable resource information on all of these areas by contacting the NACB National Headquarters.

Like AE Rho, the NACB offers a multitude of member services which are described elsewhere in this book. Like the BEA, the NACB also holds an annual national convention. The national conference is always held in Providence, and it takes place during the first or second week of November. At last count the NACB had approximately 350 member stations nationwide, and it is continuing to grow. In recent history the NACB had over 500 student station members. This total includes member stations that are of the unlicensed variety.

Considered to be one of the best member services available from the NACB is the opportunity for individual members and station members to subscribe to the NACB ListServ. This electronically-delivered medium is a vehicle by which students, faculty, advisors, managers, staff members and practically anyone can participate in an ongoing discussion about anything related to student broadcasting. The advantage of subscribing to this ListServ is that you will be monitoring discussions of the latest technological, social, economic and professional challenges encountered by actual student stations.

Anyone who is an NACB member and subscriber may participate and post any question related to student media. If you become a member and you subscribe and post a question you can expect to receive a multitude of answers. The best part of this is that there are many people who know the right answers and who are willing to help you. Membership in this organization and subscription to the NACB ListServ should also be near the top of your "to do" list as you plan your student radio station.

If you are a faculty member at an educational institution or a staff member of a radio station, you would also be well advised to subscribe to the other NACB ListServ called the "Faculty-Staff (or F/S) List." Topics discussed on the F/S list are typically in the areas of broadcast education, issues that confront student station advisors, and other related areas. The NACB also has a Faculty-Staff Advisory Board that provides advice and input to the Executive Board of the NACB.

It should be noted that the NACB was founded for student media and is still operated by students. The Executive Board of Directors consists of seven members, five of which are students who are currently enrolled in recognized educational institutions. Vacancies are filled on the board through election at the Annual National Convention. It should also be noted that past membership on the board has included both high school students and students from community colleges and two-year schools.

If you are interested in learning more about the NACB, you can call the national office directly between the hours of 10am and 6pm weekdays. The telephone number for the NACB National Headquarters is (401) 863-2225. They can provide you with membership information and answer your questions. If you have Internet access their e-mail address is nacb@brown.edu and their Web Site address is http:\\www.hofstra.edu\nacb. A quick e-mail post or a visit to the NACB web page will get you the information you need.

Many of the more successful schools are either members of AE Rho or the NACB. However, just because a school holds membership in one of these organizations doesn't mean its students are active participants. Likewise, there are many successful student radio stations which are not members or which may have never heard of these organizations. However, it should be noted that the value of access to networking opportunities, free informational resources, and the exchange of information that is available is worth far much more than the membership fees.

In addition to the two Lists that are operated by the NACB, there are many other special interest groups, or SIGs, that operate their own Lists and Web Pages. If you are interested in getting the addresses of these organizations, a good place to start is your local/school library. If your school or local library is "connected" to the net, they can show you how to use such software programs as "Web Crawler" (which is a search engine) and other available network browser programs to locate the information that you seek.

Visiting a Web Site is the fastest, most efficient method of obtaining information about an organization. It should also be noted that there are quite a few student stations that now have their own Web Sites available at the click of your mouse button. The addresses of many of these are available by contacting the aforementioned organizations.

Subscribing to a particular ListServ typically requires that you or your station become an active member in the sponsoring organization. However, you will probably find that the membership dues are minimal when compared with the wealth of available people and resources when you subscribe to a List.

Topics on a typical student media ListServ range from legal issues to engineering discussions. Because communication is very current and instantaneous, the topics usually reflect the technology. For example, recent discussions on one ListServ group have included EAS issues as well as questions about the public file and the requirements necessary for upgrading power levels of stations. For your information and enlightenment, some of these topics and issues are presented in the next chapter.

## Other Prominent Student Media Resource Organizations

There are numerous other organizations which support and address specific needs related to college broadcasting. Some organizations at the national level which focus their efforts on student media are presented in the following table.

The College Media Advisors (CMA)

Intercollegiate Broadcast System (IBS)

Special interest organizations and regional student media support groups are included in this table:

Radio TV News Directors Association (RTNDA)

National Press Photographers Association (NPPA)

The Rocky Mountain Collegiate Media Association (RMCMA)

One national organization focuses on radio and television news: the Radio TV News Directors Association (RTNDA). The RTNDA holds annual national conventions and plans annual regional conferences.

Student and professional Photojournalists can participate in annual National Press Photographers Association (NPPA) conferences and obtain certification. Some television stations have listed NPPA certification as part of the required qualifications for prospective hires. Student engineers or those interested in engineering can join the Society of Broadcast Engineers and achieve different ranks, based upon the outcome of certain tests at each level, combined with their broadcast experience.

In addition to national, regional and state organizations, consider the value of individuals who may be overlooked within your own community. In addition, please do not overlook an obvious resource person that is available to you: this author. He is more than happy and willing to help you get off to a good start. Contact information is included at the conclusion of the final chapter of this book.

# Chapter 15 ▶ ▶

## EMERGING DIGITAL AUDIO AND RADIO COMPUTER TECHNOLOGIES

*Author's Note: The decision to include a chapter of this nature in the book was not made lightly. Special thanks is due to Dr. Jerry Henderson at Central Michigan University for authoring this chapter.*

The past is analog, the future appears to be digital, and the present is a confusing conglomeration of established and new companies side-by-side scrambling for either supremacy or simply a piece of the market. New audio products, or "significant" improvements to existing ones, are brought out virtually every day. Digital audio is moving faster than the proverbial speeding bullet and leaping in gigabytes higher than the tallest imagination of only a few short years ago. One can easily remember when 25 MHz was "blazing speed" and a 20 megabyte hard drive was huge in a microcomputer. Now, as this book is being written, speed is measured in 250 and 500 MHz and higher and a 1.6 gigabyte hard drive is considered small, 2 to 4 gb is average and 9 gb or larger is wanted in broadcasting. It has been asserted that had the history of flight proceeded at the same pace as that of computers, man would have landed on the moon nine days after Kittyhawk.

Any attempt to list all manufacturers currently offering digital audio systems, whether editing, delivery or support (scheduling, logging, etc.) would be not only futile, but foolish. Not all systems available will be discussed here. What is presented as this is being written might be improved upon in many respects by the time the book is read. No recommendations of specific manufacturers or systems will be made here; but, rather, this will serve as a discussion of selected system offerings.

At the outset, it is necessary to establish some assumptions that may be helpful in this discussion:

1. Generally, since this is a highly competitive industry worldwide, prices (although designed to generate a profit) most likely will be reasonable for the product.

2. All computer platforms (operating systems) have strengths and also limitations. The reader's preference for PC (Windows) or MacIntosh is secondary to the task to be accomplished. Any bias presented by the author would be totally accidental and irrelevant.

3. Probably every system or program mentioned in this chapter will have improved with significant version changes or will have lost out in the competitive marketplace within the year.

4. Determining what will be best for a particular situation must start with a clear understanding of the specific needs to be met.

5. Reaching a final decision on which equipment to purchase will require additional research.

## How to Begin

Start with a clearly-stated need to be fulfilled and the hoped-for outcomes. A clear understanding of the need is key to ending up with what one wants. A digital broadcast system that provides "live-assist" operation is quite different from fully-automated programming with an interface to scheduling and logging software. Computer-based editing systems for stereo ("two-track") needs are less complex and less costly than multi-track production systems.

It's important to maintain the difference between "need" and "want"; for this often means a significant difference in cost, both in initial purchase as well as maintenance and/or upgrade. Therefore, although one should dream a bit and try for the best, settling for less than "all the bells and whistles" may work out better in the long run in a given situation.

Work with a budget in mind. Knowing what may be available to spend and what just isn't feasible is important not just for the initial purchase but also in determining maintenance and upgrade budgeting. What can be spent with a one-time-only capital improvement budget must be supported afterwards through the station's operating budget.

## Where to Begin

Once you have determined needs to be met and realistic budget expectations, the hard work begins. Making out a "wish list" is different from seriously researching what is available out there to fulfill your needs. Taking an appropriate amount of time to check out all options and ask the necessary questions is extremely important.

Initially, look at brochures and talk with others about what you'd like. Contact local vendors of hardware you are considering and see if a demonstration can be arranged. For computer software, check PC and MacIntosh publications (such as *MacWorld* or *PCWorld*) for reviews and comparative tests. For each and every purchase consideration, ask experts and users before committing to buy.

Especially where a large expenditure is being considered, an ideal place to go is either the National Association of Broadcasters' (NAB) Annual Convention in the spring or the NAB Radio Show in the fall. Several hundred equipment and software manufacturers and vendors are represented. One can observe demonstrations as well as try out the equipment or talk with those who already use what's being considered. Either of these two NAB events is an ideal place to narrow down options and compare price and value. Even if a purchase isn't in the immediate future, the opportunity to examine equipment and software is unequaled and sales literature is available for everything on display.

Check with the local state broadcast association for the dates of their conference(s) and whether vendors will be participating with exhibits. The National Association of College Broadcasters (NACB) and Intercollegiate Broadcasting System (IBS) conferences also may have exhibitors present.

Go on-line on the internet. Ask broadcast-related discussion groups (lists), such as that of the NACB. Look up manufacturer's information and vendor's pricing on the world wide web (www). If you don't know the Universal Resource Locator (URL), go to Yahoo or any other of the major search engines on the web and search by name or keyword.

Ask users. The fastest way to rule out an item under consideration is to hear about problems in operation or technical support from those who already have purchased it. It's also an excellent way to hear about equipment or software you hadn't considered yet.

Let's examine the issue by discussing consoles first, followed by recording media, operating systems and finally, miscellaneous comments.

## Consoles

Both analog and digital consoles are available in considerable variety. Which type is selected may depend upon many factors:

- What is the purpose for the console: On-air, production, remote broadcast, recording, sound reinforcement?

- How important are size or weight (for portability as well as fitting into limited space)?

- Is automation a factor (for station operations, for recording mixdowns, or for other reasons)?

- What equipment will be connected to it (analog or digital or both, standard broadcast or computer-based or both)?

- Is there any information besides audio that must pass through or be controlled by the console?

- What price range can be considered?

Most established manufacturers today have a full array of analog and a growing selection of digital consoles. Available brands include Alesis, Auditronics, Broadcast Electronics (BE), Dynamax/Fidelipac, LPB, Mackie, Oram, Rane, Roland, Sony, Soundcraft, Tascam, Yamaha, to mention only a few. Prices range from several hundred to several thousand dollars.

## Recording Media

Perhaps, one should start this section by renaming it "Information Storage Systems"; or given the current trend, "Digital Information Storage System" (DISS). There is little doubt that analog tape (both open reel and cartridge) is being relegated to history or for use only in limited special cases and that digital storage media are taking over the industry.

Some of the chief values of digital media are much greater dynamic range, ease and speed of operation and exactness of editing parameters. On the other hand, some significant limitations of digital at this time are different (interpret that "incompatible") operating systems, rapid change as the field develops and concern about whether a new product or manufacturer will be around in five or ten years. Any digital system being considered should be capable of interface with MIDI devices and have AES/EBU and S/PDIF I/O.

Digital Audio Tape (DAT) format has been in use for several years, first for stereo (two-track), and more recently for multi-track purposes. There are several manufacturers of stereo DAT recorders and of the standard stereo DAT tape. Multi-track DAT manufacturers include Tascam (DA-38, DA-88 with the Hi-8 format), Alesis (XT using VHS format) and Fostex. There appears to be some preference for Tascam for multi-media and post-production uses and for Alesis for home or small music recording studio purposes, though this may differ from studio to studio or in different geographical regions depending upon personal tastes. Whichever system is selected, DAT is an expandable medium (several machines can be "stacked" and "locked together") and is very stable for archiving materials.

There are some who (arguably) feel that even DAT is on the way out, to be replaced by computer hard drives, recordable CD and CD-ROM. As computers have become faster and more powerful and hard drives have increased in capacity and stability, there is a growing trend toward avoiding tape media altogether. There have been stand-alone

Digital Audio Workstations (DAW) for a few years and significant advancements in computer technology and software programming have greatly enhanced the appeal of computer-based recording and editing software. Costs have come down, also, bringing many systems within a price range affordable to smaller operations.

Several points must be evaluated when determining which recording or recording/editing system(s) to consider:

- What is the purpose for the system?

- Is there a preference for a tape or disk-based system?

- How "user-friendly" is it (how easy to understand and operate)?

- What is the training curve (how long and how hard to learn)?

- What are hardware/software requirements and costs?

- What kind of technical support will be available, how quickly and at what cost?

- What is the projected average "down-time"? (computers crash)

- Is redundancy built in to the system?

- What do other owners have to say about it?

Major audio broadcast and recording studio vendors handle a wide variety of systems. It is not uncommon for them to have active displays of several of them at their businesses. In some instances, demonstrations and try-outs may be arranged on location.

Complete stand-alone audio workstations include the *ADX Ensemble* from Pacific Research & Engineering (www.pre.com), *MicroSound* from Micro Technology Unlimited (www.mtu.com), *Sonic Studio* from Sonic Solutions (www.sonic.com), *SADiE Disk Editor* from Studio Audio Digital Equipment, Inc. (www.sadieUS.com), and *Short/cut* from 360 Systems (www.360systems.com), to mention a few.

Digital audio software for computer operating systems (PC and MacIntosh) is available generally at less cost than stand-alone workstations. However, hardware/software conflicts can arise if the computer on which the software is placed is not dedicated solely for that purpose. It is essential that the purpose and functional limitations of the specific computer be understood before software is installed and expected to work.

MacIntosh computers have been the computer of choice for music editing because of the design of the operating system. Software and software/hardware combinations for the Mac cover a wide range of capabilities, features and concomitant costs. These include the *ProTools* family from Digidesign (www.digidesign.com), *Studio Vision Pro* from Opcode (www.opcode.com), *SoundEdit16/DeckII* from Macromedia (www.macromedia.com) and *Peak* from Berkley Integrated Audio Software (www.bias-inc.com), to name a few of the better-known.

The PC platform has become considerably more useful in audio production and editing, of late. For the PC user (Windows 3.1, Windows 95/NT), some of the more well-known products include *SAW, SAW Plus* and *SAW Plus32* from Innovative Quality Software (www.iqsoft.com), *Sound Forge* from Sonic Foundry (www.sfoundry.com), *Samplitude* from SEK'D US/Soundspiration Systems (75162.1066@compuserv.com), *AudioVAULT* from Broadcast Electronics, and *Pinnacle Project Studio* from Turtle Beach (www.tbeach.com), among others.

## Operating Systems

There are several radio station operating systems available, most of them PC DOS or Windows based. In trying to decide what system to buy, it is imperative to detail very carefully exactly what is desired from such a system. Costs vary by several thousands of dollars and many systems can be custom designed to meet specific needs, with the price proportionate to the complexity and specifications of the system. Technical support, and also redundancy, become matters of major concern for this kind of system since care must be taken to assure ability to remain on the air and minimize down-time when problems occur.

Systems range in sophistication from so-called digital cart machines to complex multi-station LAN-based operating systems with integrated scheduling, billing and logging capabilities. The following represent this broad range of price and sophistication: *DADPro* from ENCO America (www.enco.com), *Master Control* from Radio Computing Services (www.rcsworks.com), *Scott Systems* from Scott Studios (www.scottstudios.com), and *The Digital Jukebox* from Digital Jukebox (www.digitaljukebox.com).

## Miscellaneous

In conjunction with digital operating systems, an abundance of music and spot scheduling, billing and logging systems is available. One of the most well-known and widely used is *Selector* from Radio Computing Services (www.rcsworks.com). Another, dealing especially with billing logging, is *Traffic Master* from Register Data Systems of Perry, GA.

Don't forget to plan and budget for furniture. Old, scratched, marred tables and racks can totally detract from the perceived value of that new major equipment investment and make it look instantly shabby. Well-designed new standard cabinets, racks and desks can be found in catalogs of most equipment vendors. In shopping for studio furniture be sure to shop around as prices and features may vary dramatically. Also, custom furniture can be ordered from a multitude of cabinetmakers. As with anything that is custom, expect to pay a premium price for this kind of specialized carpentry work.

Another aspect of radio today is *Webcasting*. Initially, the station operator must determine whether reaching a new, diverse and widely-scattered listenership is as important as focusing on the local area for which the license was obtained . If proceeding with a streamed audio presence on the world wide web is desired, decisions will have to be made concerning operating platform, server(s), amount of RAM, connection size, website administration, maintenance and a host of other concerns which should be based upon expected visits per day from internet customers. Full discussion of the merits of such systems and how to construct them can be found at many locations on the internet. It is important, here, mainly to stress that the webcasting system would best be kept separate from the signal flow of the main broadcast system. Although broadcast and webcast systems can operate compatibly, they are designed to serve different purposes and should be administered and maintained separately.

As in most anything, one gets what one pays for and one pays for what one gets. Whether a basic, relatively simple system is needed or a fully-integrated sophisticated operation is desired, careful planning, attention to detail and deliberate, focused research is absolutely critical. One of the greatest joys is watching it all come together and work the way it was hoped; one of the most frustrating moments is if one discovers that what was purchased wasn't what was needed. Corrections, changes and last-minute add-ons can destroy the budget as well as cause undesirable management outcomes.

## Summary

Choose your equipment purchases carefully after researching your final selections as painstakingly as time and energy permit. Network with friends, acquaintances and professionals and collect as much information as possible. Take your budget into careful consideration and be sure you are getting the best value for your investment and that the equipment will have the capacity to grow with the station's future needs.

# Appendix A ▶ ▶

## COMMONLY ASKED QUESTIONS AND FCC RULES THAT MAY APPLY TO YOU AND YOUR STATION

Many advisors are thrust into the challenge of overseeing a student radio station or are given the role of helping a group of students get a station started. Likewise, church leaders or a civic group starting an NCE radio station may not know the first thing about laying the groundwork to get the project started.

There are numerous questions and where to turn for the right answers may be baffling. This book is intended to suggest some places to start looking for answers.

This section presents some of the commonly asked questions coming from beginning student radio broadcasters and others who want to start a noncommercial radio station. Some of the most challenging of these involve FCC rules and regulations.

This chapter cannot answer all of these, but it will address many of the commonly asked questions that are topics of current discussion and concern. Some of the following questions and answers were electronically posted on the National Association of College Broadcasters' (NACB) ListServ. They are reprinted here with permission of both the NACB and the individual authors.

### What Are the FCC Rules that Pertain to Unlicensed, "Part 15" Stations?

The following answer comes from John Devecka at LPB, Inc.:

AM regs allow :-100mW input power to an antenna and ground lead combination with a maximum length of 3 meters. These systems will usually produce a 400 ft radius or so, depending on clarity of your local spectrum. Stations running miles with this equipment are generally illegally grounding in order to radiate their support structure (such as a billboard).

Any input power to an electrical line in a Carrier Current application such that your radiating field when measured at X meters from the power lines does not exceed 15 microvolts per meter. X = 47,715/frequency in kHz. Low end is best, typically reaching a distance of 50–150 feet from electrical lines

There is an additional minor change for educational facilities to use a vertical antenna within their perimeter, but the field is too low to be of use WHEN DONE WITHIN THE REGS, which 99% of those I have seen are not.

Radiating cable systems (thanks to LPB—shameless plug) are also allowed to operate within the same formula as the Carrier Current applications, just measured from the cable instead of power lines. Typically 100–200 foot reception.

FM regulations are restricted to 250 microvolts per meter field strength at a distance of 3 meters from the antenna. Additionally, the antenna must be a permanent part of the device, not replaceable with a *bigger* antenna. This is the ONLY FM rule for power. There are no 100mW FM regs. FM is a shambles in some areas and 250 microvolts is useless, but in other areas, with some decent height you can cover hundreds of feet— especially in MONO.

Note that in ALL of these cases you must be using a transmission device which is FCC Part 15 Certified, and bears the ID number to prove that. There are PLENTY of illegal kits and finished devices out there, but precious few actual manufacturers with actual engineers and assemblers building legal broadcast transmitters. Every so often the FCC notices that.

Any other questions on Low Power Broadcasting <ahem> I mean LEGAL Low Power Broadcasting?
John Devecka, LPB, Inc. LPBINC@aol.com

*The following is a response to the above statement:*

What is the FCC's "official" response on the issue of illegal, unlicensed "pirate" broadcasting?:

I really appreciate the posting of this information. I often receive calls from individuals (students and others) who are bound and determined to operate a broadcast station. If they can't get it legally, they will operate illegally. I want to emphasize that the latter should never be considered. If caught with an illegal station by the Commission, the Commission will seize the illegal equipment and turn the operators over to the Justice Department for prosecution. Conviction is a Federal offense and I am a firm believer in squashing this type of thing. You say that the station is close to your position on the radio dial, while not causing interference, may be hurting your signal in some other way. Perhaps this will encourage other pirates to start up farther away from your community that will hurt your signal. In other words, while initially these guys may be OK, they may down the road hurt your efforts and those of other stations. I would prepare and file a complaint with the FCC field office nearest you. In fact, you may want to solicit other stations in your area to this question.

If the pirate station starts selling ads, then this cuts into their operations as well. As for their claim that their stations do not interfere with existing stations by listening to blank spaces on the dial, they do not have a clue many times as to adjacent channel interface. Their operation may effect the reception of a legal station in your area.

Sorry for the rant, but you hit a sore spot.

I am not a fan of FRB (Free Radio Berkeley) or its tactics. In no way has the constitutionality of what they are doing been decided in their favor. In fact, the FCC is going ahead in the court cases. To date, FRB has been fined tens of thousands of dollars for illegal operation. I disagree with his tactics, which encourages the possible radio chaos that originally spawned the FCC. Broadcasting is not a right, as it is regulated. I was not very happy with the FCC's decision to get rid of 10 watt stations or at the very least demote them about 15 years ago. However, what some of these kinds of companies are doing will make it impossible for any legitimate reform in this area or rule changes to restore that service.

## What Is the Relationship of Tower Height to Power Output on FM?

A100 watt station on a 1600 foot mountain or tower, talks better than a 1000 watt station on a roof top. Which leads me to the question of the moment: How do you tell people audience reach?

We face this issue frequently. Our ERP is 8.5 kW, with a HAAT of 716'. Because of our unusually tall tower, power alone does not accurately reflect our coverage. For what good it does, I always report height along with power—even if I'm not asked. It probably doesn't mean anything, but at least I feel better.

I often will also report the distance to and population under our 1mV/M (60 dBu). If I'm really testy, I'll also give the .5 mV/M (54 dBu); I think it's a more realistic, yet still conservative, indication of a "listenable" range. Whenever I give contour distances, I always label them (whether the labels are meaningful is another issue). One local commercial station (100 kW, 1650' HAAT) shows their 34 dBu contour in their sales literature! They're not lying about their coverage, since the map is labeled accurately!

Our terrain is flat enough that our pattern is very nearly circular, so averaging the distances to the contour doesn't make a big difference. For folks with "significant terrain features" this would be a bigger deal.

I have never understood why people (i.e., record companies) are so fixated on power alone, when distance to the 1mV/M is something that every broadcast station must compute, and would be an easy way to compare apples to apples. Although not necessarily a true indication of "listenable" area, at least the 1 mV/m is objectively defined.

My predecessor used to report (even in the *Broadcasting Yearbook*) our ERP as 5kW, even though the (expired) CP read 4.265 kW and the license read 10 W TPO (oops, that's another long story). I never thought this "rounding" was ethical, let alone legal.

"Broadcasting with the effective power of 100 zillion watts (at 10 feet) . . ." telling industry people that they are "20,000", as that is their "effective" power. Not quite so honest, since their ERP is only the 3KW. In addition, I assume that this could get you into some ethical questions, or the possible problem situation if you are "caught" by some knowledgeable person.

The college station I worked at had an ERP of 235 watts, but the TX sits atop a mountain. Because of this, it has a radius of 65 miles or more. When record labels or potential underwriters asked about our wattage, I used to say "We're 235 watts, BUT because of the transmitter height, we have a 65-mi radius and a potential audience of 300,000." Once it was clear what our REACH was, people were more interested in doing business with the station.

(4 yrs. in Student Radio)

## Engineering and Little Known Applicable FCC Rules

To know all of the rules that can apply to noncommercial educational radio stations, you had either be a communications attorney or be familiar with CFR 47 Parts 70 to 79, particularly Part 73.

For example, there are a sizable number of NCEs that sign off after being on the air only a limited number of hours per day or limited hours per week (for example, 5 hours per day Monday through Friday). Little do these station operators know that they are in violation of CFR Part 73.561a "Operating schedule; time sharing" which states in part that "All noncommercial educational FM stations are required to operate at least 36 hours per week, consisting of at least 5 hours of operation per day on at least 6 days of the week, consisting of at least 5 hours of operation per day on at least 6 days of the week."

It is comforting to bonafide educational institutions that the rule goes on to state that: "however, stations licensed to educational institutions are not required to operate on Saturday or Sunday or to observe the minimum operating requirements during those days designated on the official school calendar as vacation or recess periods."

Of special interest to all stations is part b of that same rule: "All stations, including those meeting the requirements of paragraph (a) of this section, but which do not operate 12 hours per day each day of the year, will be required to share use of the frequency upon the grant of an appropriate application proposing such share time arrangement."

Tell us about underwriting. What is legal, and what constitutes an illegal underwriting announcement? Can "full blown" commercials be sold and aired legally for non-profit corporations?

#### #20 NEWMAIL

Subj: underwriting
*Is underwriting considered a tax deductible donation?*

#### #21   NEWMAIL

Subj: REPLY to underwriting question
Yes, in MOST cases, underwriting is a tax-deductible donation. Here are some things to consider:

1. Is your station, or its owners (e.g. University Board of Trustees) noncommercial by tax definition (501 c3, etc.)?

2. Are we talking about personal or business donation?

(a) If personal, check your state laws concerning any limits, whether financial or to what type of organization.

(b) If business, the business entity may want to check with their accountant(s) to determine whether they have reached the $ limit they can "donate" AND ALSO as to whether it would be better for the business to label it as a tax-deductible donation or as an advertising expenditure.

(Don't confuse the limits of your donor announcement with how they list their donation. They can call it what they want, YOU have to follow FCC rules concerning announcements.)

3. Do you have a "Grants" Agreement form for the business to sign which specifies that they are:

(a) Giving the money or gift-in-kind (e.g. CDs from the record shop) as a donation to your station;

(b) Recognizing that this is a donation and that you (the station) will make donor announcements—but are not guaranteeing a specified number of announcements; and

(c) Understanding the limitations of what you can and cannot say or do (stylewise) on the air.

The main reason for this Agreement ("contract," if you will) is not legal but is for up-front clarification of understanding between you and the business as to what this is all about. Over the many years, it is my unavoidable experience that 99% of all businesses do not understand the difference between advertising on the local commercial station and on your noncommercial station and are upset when they don't hear a full-blown 60-second CA at least "x" number of times for the donation they made.

Far better to lose a few donors than to have ill will and bad PR running through your territory.

Business donors are often tough to get and keep—but well worth the effort if you target appropriately for your listener.

BEST OF LUCK!

*#22*

It most certainly is. Generally, an underwriting "buy" is just like a charitable donation to a non-profit organization, which is tax deductible. Your station's on-air announcements merely acknowledge the support. And, as far as I know, advertising on commercial stations is also tax-deductible since it's a business expense.

Hope this helps.

*#23    NEWMAIL*

Subj: underwriting -Reply
Let me be the first, of probably many, to say that the answer is "yes".
Allen Myers (amyers@fcc.gov)
>>>Is underwriting considered a tax deductible donation?

*#24Subj: Underwriting: churches*

As long as we are discussing underwriting, are there any specific rules regarding underwriting paid for by religious institutions?

We have a non-denominational black gospel show that wants to begin soliciting contributions, and it looks as though it will be very successful.

"There is absolutely no inevitability as long as there is a willingness to contemplate what is happening."—Marshall McLuhan

*#25    NEWMAIL*

Subj: Underwriting: churches -Reply
This message warrants some discussion. As long as the sponsoring organization is a bonafide non-profit organization, Brian can air its solicitations like he would for any other charitable organization—pro-

vided that the solicitations do not interrupt regular programming. [From Brian's post, it appears that the solicitations will be made in the course of the show, so this may not be a problem.] If there is to be a problem, it will come in Brian's determination which programs to air with this type of announcement. If he tries to permit some organizations to do this and not others, he might receive objections from the organizations which he has denied. Therefore, my recommendation would be that he establish a clear station policy on this issue that organizations must adhere to before he accepts their programming.—Allen Myers, amyers@fcc.gov (202) 418-2774

*Subject: Re: Underwriting*

> Okay, as we get detailed with underwriting, can the following terms be used?
> 'variety'
The word itself is not qualitative, but it would depend on the context. Something like "Support for KXXX is provided by Mom & Dad's Pretty Good Hardware, offering a variety of power tools for professional use," is probably o.k. Saying something like "has the biggest variety" would be against the rules.
> 'much more'
Red Alert . . . Shields up Mr. Worf.
> Additionally, can slogans be used, even if they imply some quality Slogans can be used, even if they are qualitative, IF it can be shown that these slogans are a long—time part of the corporate identity. "GE . . . we make good things for life." "Smuckers . . . with a name like Smuckers . . . it has to be good." These are examples of slogans which, though qualitative per se, are legal because they are an integral part of that corporation's public identity.

*Subject: Re: Underwriting*

. . . and for the most part, they are copyrighted slogans, and therefore can be used. The safe option is always to skip the slogans!!

*Subject: Re: Underwriting*

At a recent NACB National, Attorney Carey Tepper again handed out a booklet with common do's and don'ts. The logo phrase example he used as being okay was GE—we bring good things to life. The example used as being illegal was "Get Met-It Pays". I'm sure the NACB office can get you copy of the handout if you call.

*Subj: FCC License Requirements*

I am an Operations Manager and Chief Operator. I am writing to anyone who can offer a bit of wisdom on the absolute license requirements for FM stations chief operators and operations managers. I understand deregulation has made the Radiotelephone Operator Permit or "RP" unnecessary. My question is this: Am I as chief operator and operations manager required to hold any kind of license? I am not certified by a broadcast-related engineering society but I have been interviewed and selected for this position and deemed competent by our licensee and I do rely on regular help from the contracted services of professional radio engineers. Any information regarding licensing requirements is greatly appreciated.

*Subj: FCC License Requirements—Reply*

The Commission no longer requires that chief operators hold a license. The individual licensee is responsible for employing and training a chief operator who is capable of keeping the station in compliance with the Commission's technical rules and regulations.

Allen Myers, amyers@fcc.gov (202) 418-2774

*Subj: Tower Light Observations*

Are we required to make a log of our tower light observations?

In the past we have made tower light observations daily and then logged them into our station logs. We currently make the observations but they are simply noted at the bottom of our log sheets. Are we required to make specific record?

Also, in the past we have received an FM Broadcast Station Self-Inspection Checklist (Bulletin CIB-18FM June 1995) that we have used to keep up to schedule with these regulations. Is there an updated version of this checklist that we can get a hold of? If so, do all of the rules apply to noncommercial, educational college radio?

Hope these questions are clear thanks in advance for the help,

*Subj: TOWER LIGHT OBSERVATIONS -Reply*

Please follow the information contained in the "FM Broadcast Station Self-Inspection Check List". It is the most current material we have available.

—Allen Myers, amyers@fcc.gov (202) 418-2774

*Subj: Remotes*

At our FM noncommercial station we've worked out a deal with a local cellular company. They lend us a phone and give us airtime and we run an underwriting spot at the beginning and the end of each break, then we'd also run a few additional spots throughout the week. Not a bad deal!

All we used was the phone. Our engineer rigged up the patch to the board that let us broadcast the phone signal (a fun project for the whole family!). Our remote people we just talent in the field (and banners and stuff). The music came from the studio. The DJ would fill the Talent in and vice versa.

## Subj: RE: Remotes

> I mentioned this in my last list message: mobile broadcasting.

> Again informing the list of how little I know, what are the options?

> I want to be able to do about 6 mobile broadcasts of a few hours a year, and really do not need to send the music from the mobile location.

> I have been told that we could simply hook a cellular phone with a small mixer up to a few mics, and with an operator back at the studio, broadcast "live" from the event. Will this work, and will the voice quality carry over?

> I also have heard a little about mobile transmitter units? What do these run, and how do the operate (FM, shortwave, etc.)?

> I am interested in securing a mobile option soon. Thanks again for your answers to previous questions.

>Cellular "lines" have improved in quality over the years, but you still have very limited frequency response. The sound of cell lines is acceptable for voice-only remotes (like sports and news), but music would sound terrible.

Depending on the frequency congestion in your area, radio Remote Pickup Units might be a possibility. Equipment costs will be about $2500–3k. We use two such systems, one (~455 MHz) for our program itself and another (~161 MHz) for two-way, off-air communication between the studio and remote site. Systems in the 450 MHz region are good for about 10 kHz frequency response (there are actually two 100 kHz wide channels that you can get 15 kHz out of). Keep in mind that a single system will only allow you mono transmission. Stereo doubles your equipment and cost. Frequencies are shared between stations, so we use switchable dual-frequency units to be sure we have a channel. RPUs must be frequency-coordinated and licensed.

Another low/no cost solution you might look at might be your campus phone system. If your university owns the system, you might be able to get lines at little or no charge for broadcasts around your own campus. I was able to get our university phone guy to permanently punch through a couple of dry pairs directly from campus locations to our studio. With a little engineering work to balance out and equalize the lines, we have been able to do high quality, stereo music broadcasts at absolutely no cost. This is one example of how it pays off to politick around campus; I've been cultivating the phone office "connection" for many years. The phone guy likes us; he decided it wasn't worth the effort to put through the paperwork to charge us for his work, or for the lines. In these parts they refer to it as the "good ole boy" system.

## Subj: Danger with underwriting services

>I might also mention another company which markets underwriting for PSA-type messages. Perhaps there are similar firms. That can help tremendously.

>The issue with companies like this is, in order to be profitable, the price they charge outside advertisers for your airtime is double or triple your normal underwriting rates. This can be a problem when one of your station's student sales staffers goes to a local business/merchant, offering underwriting at a given rate, and then the merchant gets pissed off. That's because they were called by the underwriting services company your station enlisted, they bought into it, but paid triple the price that your student is now offering for the same thing! If this gets around (local merchants talk to each other a lot), it ruins your reputation mighty fast, and then no business wants to underwrite on your station.

The way to avoid this is to specify a certain territory for the underwriting services company to cover (they will agree to this IF you insist) which won't overlap with the territory that your student sales staffers-in-training want to do. Typically, that means keeping those within walking distance of campus for yourself, plus a few key accounts outside the territory (specify them by name in your agreement with the underwriting services company). The rest is theirs to pitch. Good luck!

## Subj: RE: Danger with underwriting services

Good advice. I could not have put it more succinctly! Having done Commercial Radio Sales, I can say that a lot of Commercial stations avoid these "boiler room" marketing agencies like the plague. Others follow the previous suggestion and still others fall for their line and get burned by earning a bad reputation. Another tip is to talk to the marketing agencies references and ask them about the issues raised on this thread.

A previous post is very true, as ham equipment does not have the duty cycle for continuous broadcast. However, there are two other concerns more important. You legally can not use Amateur frequencies for broadcast. This is a big No No!

In addition, ham equipment is not type accepted for use in Part 74 applications, or for that matter, for use in business communications. So even a radio that is valid for ham, and might be modified to transmit outside the ham frequencies, can not be used for broadcast purposes.

By the way, the same is true for any Part 15 broadcast applications. You can not just take an old Part 73 exciter and stick in on the air. It needs to be type accepted for Part 15.

### Subj: Digital v. Analog

This has been on my mind for a while now. What is the difference between Digital and Analog? This is more in reference to a radio station itself being Analog or Digital.

### Subj: REPLY TO: Digital v. Analog

Same as the difference between a CD and an LP. The analogy between a CD and an LP is not incorrect but there is still a missing part. One of the main advantages to true digital compared the analog, is that analog is linear, whereas true digital does not have to be linear at all. Analog is a continual sine-wave which produces a specific frequency, where digital is a binary representation of a certain analog signal. The digital allows for compression to occur without noticeably degrading the signal. You can go too however with compression. The toughest factor to make the switch to digital audio is the storage devices that must be invested in. Hard drives are coming down in cost, but for a true digital transportable medium, tape is still the cheapest way to go. That's my two cents.

No matter how much we try to go to digital and non-linear forms of editing, the end result is still viewed in a linear fashion, until that changes, digital may not be the practical, efficient systems. What I meant to hear was what is the difference in equipment, sound, and power for a digital station vs. an analog station.

## Concerts

hi, I just wanted to ask a question about our radio station promoting & sponsoring small concerts and shows. I remember hearing something at the national about the subject as related to underwriting and what can be said over the air to promote such events.

Does the language used & the actual text read over the air depend totally on our amount of involvement (money-wise) with event depend on what can be said? right now we rent out our portable sound equip. for small shows and usually get a decent rental fee, but we also promote all sorts of musical events that we don't do the sound for. there's a variety of places (bars, clubs, park shelters, motels) and door charges involved.

Its not that we don't mind promoting all the musical events in town, but I just wanted to know what was legit as far as FCC/underwriting/promoting, and the language used.

thanx for any help.

*Subj: promoting "concerts" -Reply*

Received: by gatekeeper.fcc.gov; id;

From: Allen Myers <AMYERS@FCC.GOV>

Pursuant to Section 73.503 of the Commission's rules and regulations, no promotional announcement for a concert sponsored by a for-profit organization may be made. Promotions for concerts given by non-profit organizations are acceptable.

Allen Myers (amyers@fcc.gov) (202) 418-2774

I mentioned this at a panel I did with Gary Hawke on station promotion at the National NACB conference, so I'm probably guilty of raising Jeremy's question. My comment was based on a 1983 NAB "Counsel from the Legal Department" memo, that reads:

". . . The purpose of this Counsel Memo is to alert broadcasters to the pitfalls to be avoided in dealing with the promotion of music concerts which may or may not be sponsored by the broadcaster. The urge to promote a concert with the lead-in that "Station ROCK presents . . ." may not only be tempting, but it may also be misleading in the view of the Commission and could result in substantial fines or other sanctions.

To this point, the Commission has had infrequent opportunity to express its views on station promotion of concerts, but it is safe to say that there are important guidelines for the broadcaster to follow in formulating its concert promotional announcements. These may be summarized as follows:

(a)    Where the station has made no financial investment in the concert and has not devoted an effort to making the show a success, don't announce that 'WXXX presents . . .'

(b)    If the station is actively involved in arranging a concert, this commitment, which may involve only limited financial investment, will probably support a station announcement that it is at least a co-promoter of the concert.

These guidelines encourage a careful analysis of each promotion by the station's staff. The following factual examples discussed by the Commission in the few cases available to the public may be helpful in permitting the station staff to avoid an unwise and perhaps costly decision.

In one case, the Commission found that a station had broadcast misleading announcements regarding the source of a concert's sponsorship by announcing that "KITE presents' the Tony Bennett concert when, in fact:

(a)  the concert arrangements had been made by other parties pursuant to contract;

(b)  KITE had no financial interest in the concert;

(c)  KITE assumed no responsibility for production or management;

(d)  KITE did not receive a percentage of the gross revenues from the concert; and

(e)  KITE merely publicized the concert on behalf of the actual sponsors in exchange for tickets and promotional privileges.

The KITE case followed the Commission's analysis in one earlier case where the Commission found that a substantial question had been raised as to whether a station's announcements were misleading where the station assumed no responsibility for production or arrangements but simply received one percent of the gross revenue.

If, however, the station is actively involved in arranging the concert no financial commitment is needed to co-promote the show. For example, WDAF was allowed to represent itself to the public as a co-promoter when it did the following:

- helped with the selection of concert performers;

- assisted in arranging ticket outlets;

- gave advice on the concert's sound system, lighting and/or staging;

- gave suggestions about promotional appearances by concert talent in support of the concert;

- assisted in negotiating talent, support personnel and facilities fees;

- used station personnel to assist in the coordination of various elements of the production;

- gave advice about an appropriate hall for the performance;

- presented live broadcasts of some concerts it co-promoted; and

- organized, or helped to organize, related events preceding of following a concert in some instances.

Although WDAF's degree of involvement was sufficient to allow it to co-promote the show, there is no magic formula for determining minimum amount of responsibility and/or financial investment necessary. Each situation will depend on its own facts, but should be discussed thoroughly by appropriate station personnel or with communications council.

In sum, although limited in number, the few available Commission pronouncements regarding concert promotion should give the careful broadcaster a basis for applying common sense to its copy writing. Before you announce 'WXXX presents . . .' ask: what is the station's actual position?

# Appendix B ▶ ▶

## THE ISSUE OF LEGAL PIRATE RADIO STATIONS:

### INTENTIONAL ABUSES OF CFR PART 15

---

## Permitted Forms of Non-licensed Low Power Broadcasting

If you are interested in starting a student radio station and getting it on the air quickly, you may wish to consider this route. Part 15 of the CFR does outline the rules which regulate low-power, unlicensed transmitters. In fact, you may decide to build a low-power radio station and operate it while you undergo the project of preparing and filing an application for a construction permit for a licensed radio station.

It is recommended that you write to the FCC and request OET Bulletin No. 63 titled: Understanding the FCC Regulations For Low-Power, Non-Licensed Transmitters (Edited and Reprinted February, 1996). There is also another document which has been prepared by the FCC titled: "Information About Low Power Broadcast Radio Stations."

This information is available on the Internet at the FCC's Web Site address: http://www.fcc.gov/mmb.html under the Audio Services Division. At this address you can also find the document titled How To Apply For A Broadcast Station which can likewise prove to be useful in your endeavors. If you do not have access to the internet, you can call the Audio Services Division at (202) 418-2720 or Consumer Assistance at (202) 418-0190.

## Part 15 Devices

According to the Commission's flyer titled *Low Power Broadcast Radio Stations:* "Unlicensed operation on the AM and FM radio broadcast bands is permitted for some extremely low powered devices covered under Part 15 of the FCC's rules. On FM frequencies, these devices are limited to an effective service range of 35 to 100 feet (or 11 to 30 meters). For the exact wording of the rule, please refer to 47 CFR Section 15.239. On the AM broadcast band, these devices are limited to an effective service e range of approximately 200 to 250 feet (61 to 76 meters). See 47 CFR Sections 15.207, 15.219, 15.209, and 15.221. These devices must accept any interference caused by any other operation, which may further limit the effective service range. For more information on Part 15 devices, you are invited to contact the Consumer Service Branch, Office of Engineering and Technology at our Columbia, MD office, phone (301) 725-1585 extension 210."

## Prohibited Forms of Low Power Operation

Again quoting from the FCC's information sheet titled *Low Power Broadcast Radio Stations:* "A license or a construction permit is required for forms of operation in the AM and FM radio broadcast bands which cannot be classified as Part 15 or Carrier Current Stations (also referred to as Campus Radio Stations). This license or construction permit must be obtained from the Commission prior to construction of a broadcast station and before operations may commence. For more information, refer to the FCC's information sheet titled *How to Apply for a Broadcast Station.*"

## Penalties for Operation Without a Permit or License

According to the Commission's flyer titled *Low Power Broadcast Radio Stations:* "The Commission considers unauthorized broadcast operation to be a serious mater. Presently, the base penalty for operating an unlicensed or "pirate" broadcast station (one which is not permitted under Part 15 or is not a Carrier Current Station is set at $10,000 for a single violation or a single day of operation, up to a maximum amount of $75,000.

Adjustments may be made upwards or downwards depending on the circumstances involved. Equipment used for an unauthorized operation may also be confiscated. There are also criminal penalties (fine and/or imprisonment) for "willfully and knowingly" operating a radio station without a license. DON'T DO IT!"

Despite the FCC's warning, it is estimated that there are between 350 and 400 unlicensed, illegal "pirate" radio stations presently operating in the United States. Some of these pirate radio station operators are so brazen, brash, blatant and open about their activities as to advertise and promote their frequencies and broadcast hours in the local newspapers and to sell advertising to area businesses.

## Free Speech vs. Their Right to Broadcast

There are those who support the argument of Free Speech vs. their Right To Broadcast. According to page three of the Commission's flyer titled *Low Power Broadcast Radio Stations:* "A number of inquiries received at the Commission are from persons or groups who believe that there is a First Amendment, constitutionally protected right to broadcast. However, in such cases as National Broadcasting Co. v. United States, 319 U.S. 190 (1943), the Supreme Court stated: '. . . The right of free speech does not include, however, the right to use the facilities of radio without license. . . . Denial of a station license on that ground, if valid under the Communications Act of 1934, is not a denial of free speech.'"

## Stephen Dunifer and Radio Free Berkeley

In what has become the battle cry for pirate radio station operators nationwide, Stephen Dunifer won his case in a California Federal Court. Dunifer, a broadcast engineer, operates a low power station called Radio Free Berkeley. The case involves an interpretation of the FCC rules (47 CFR Section 73.211) which sets the minimum power for an FM licensed broadcast station at 100 watts.

This rule has been interpreted by some individuals who believe that stations less than 100 watts do not need to be licensed. Some broadcast practitioners believe that if there was ever any hope of resurrecting the old Class D rules which permitted operation of licensed 10 watt stations, the RFE supporters have ended that possibility.

In the interim period, the US government has appealed the decision in the RFE case. Until the appeal is heard, many new pirate operators have gone on the air in the hopes that the RFE decision will be upheld and anyone will be able to operate an over-the-air unlicensed FM station at a power level less than 100 watts.

## Is Less Than 100 Watts FM Legal if You Don't Interfere with Another Station?: Radio Free Lenawee

According to Pastor Rick Strawcutter of Adrian, Michigan, "a 'pirate' infers that you are doing something wrong. Strawcutter is operating an unlicensed 95-watt FM stereo radio station which he calls Radio Free Lenawee, named after the county in which the station is operating. He has also referred to the operation as the "Rosa Parks of radio."

Strawcutter claims his 95-watt unlicensed FM radio station cannot be inspected by the Federal Communications Commission because (in his words) "they do not have the authority to inspect an unlicensed station." When representatives of the FCC attempted to inspect the station, they were handed a paperwork package put together by Strawcutter's attorneys, Constitutional Litigation Associates of Detroit, and turned away by Strawcutter, who said the representatives were unable to cite their authority for coming to inspect the facilities (*Radio World,* Jan. 8, 1997. p. 7).

FCC Spokesman John Wilson said the case of Radio Free Lenawee remains under investigation and will be thoroughly investigated. "Operating a radio station without a license obtained through proper procedures required by the FCC can result in a fine of up to $10,000 and confiscation of all equipment being used to broadcast," according to Wilson.

Pastor Strawcutter maintains that what he is doing is "absolutely constitutional and lawful." He cited the California Federal Court case involving Stephen Dunifer, a broadcast engineer who runs a low power station called Radio Free Berkeley.

"The court essentially ruled that the FCC's current ban on low-power broadcasting—that is anything below 100 watts—is not the least restrictive means," said Strawcutter. "When any bureaucracy is charged with regulating a constitutionally protected right, such as speech, they must do it in the least restrictive means in order to balance the people's right to be able to do something, against other people's rights not to have something interfered with."

Strawcutter claims that the FCC's refusal to even consider licensing anything under the 100 watt level is too restrictive and not constitutional. Strawcutter claimed that he does not have a bone to pick with the FCC as a regulator.

Strawcutter believes that we need a smooth transition of traffic in the airwaves. He continued: "But as with so many bureaucracies, we're getting more and more regulated and controlled in our society. The problem arises when someone wants to get access to the airwaves. They're generally faced with all sorts of bureaucratic haggling and hoops to jump through."

Coincidentally, FCC Chairman Reed Hundt has declared "a pro-competitive, deregulatory national policy framework" to be in the mantra of the commission for the coming year. Whether that deregulatory trend would benefit a low-power broadcaster like Strawcutter is far from clear.

According to an article in the January 8, 1997 issue of *Radio World* magazine: Radio Free Lenawee is on the air 24 hours a day at 97.7 MHz. Volunteers run the morning show from 6-9 am followed by an hour segment with Strawcutter and then more live programming again at noon. Satellite programs such as The American Freedom Network fill the remaining slots.

The 95 watt signal broadcasts solid coverage for six miles, and even up to 15 or 20 miles, depending on how good your radio is, according to Strawcutter, who maintains that the signal doesn't interfere with any other licensed station.

He contracted an engineer to determine the frequency and went to what he claimed was "tremendous expense" to get a type-accepted transmitter and antenna system.

"I could have used transmission line for 50 cents a foot,

but I paid $3 a foot for high-quality transmission line. I've got a QEI 695 exciter, the amplifier is a Bext PF250 and the antenna is a Celwave."

"All this," claimed Strawcutter, "is my way of narrowing the controversial issues to one question: Do I or don't I have the right to broadcast?"

Strawcutter's station kicked off operations November 4, 1996 and estimated an audience of between 15 and 20 thousand the first day. We had 250 telephone calls live on the air and the next day was the same, according to Strawcutter.

Strawcutter said he is not faxed in the least by the FCC's threats to shut him down. He said he is not standing in the defense of everybody who broadcasts at low power because "some of these transmitters are causing harmful harmonics and spurious radiation and do cause interference." Still, he said he knows that he is right.

"They know that I'm right," he said, "and they (the FCC) should step aside. This is the '90s! The internet. You can sit down at a computer and virtually communicate with the world.

So what's the big deal about a peanut-powered radio station communicating with a few thousand people?"

Strawcutter predicts that within a year 500 to 1000 new low power radio stations may pop up around the country. He claimed to have put the station on he air for $10,000, but said he could have done it for $5,000.

Strawcutter's attorney, Pat Edwards, said First Amendment rights are at stake. "The FCC may argue that there's some compelling reason that they have to restrict less than 100 watt, but right now, they haven't come up with that argument," he said.

The next step rests with the FCC, said Edwards, who predicted a long an drawn-out process to come. The FCC has appealed the RFE case. In the meantime, the pirate radio station operators across the country seem to be sighing a breath of relief for the moment.

## NACB ListServ Comment Corner

What follows are comments dealing with CFR Part 15 violations taken from the NACB ListServe with permission of the authors and the NACB.

> From: John Devecka <LPBINC@AOL.COM>
> Subject: Re: AM BROADCAST

> In a message dated 07-24 18:36:50 EDT, you write:
> << Goodday from KRUI-FM in Iowa City, Iowa. I would like to post a question concerning low wattage AM broadcasting. Our station is looking into the addition of a training wing that will also serve in better serving the community with more news and sports coverage.
> What do we need to consider when looking into this proposal? We plan to cover the campus of the University of Iowa which, for our purposes, is about a 5 mile radius. How do we select frequency and call letters? If anyone has been in a similar situation, please respond.
> —thanks, operations director >>

As Michael Black mentions, your campus layout will greatly dictate the type of AM system which can be used on the campus for open air broadcasting. Both carrier current and antenna-based systems can

work for the school, depending on a number of issues. Please e-mail me with your snail mail address and I will send you about 40 pages of details on AM carrier current rules, design and theory.

Note that while covering the community at large is a great idea, it is not permitted without a license and would be considered pirate radio by the FCC. Your field strength is either limited at a distance from the buildings (in the case of carrier current) or at the perimeter of the campus (in the case of an antenna system). In both cases the field strength is based on the frequency you have selected. Antenna systems are weighted to the top end of the band (1610kHz+) and carrier current to the lower end (530kHz). While there are companies that will sell you product to reach the radius your seeking, you are the one responsible for the end result and field verification of your compliance.

You might consider working with the local CATV company to reach your off-campus target audience as well. While CaFM has some drawbacks, it can be very inexpensive and would provide a wide area of reach for the station.

You also have the option of purchasing one of the many AM radio stations in your area, some of which may be off-air now and available. The list of stations that are currently considered "dark" by the FCC is located at:

http://www.fcc.gov/mmb/asd/amsilent.txt

As for the selection of frequency and call letters for the station, you would be operating under Part 15 as an unlicensed service, there are only a few restrictions. The call letters can be anything you want, but should not "sound" like another station in the market, just to avoid confusion. The frequency must (1) not provide harmful interference to a licensed broadcaster, (2) must accept any interference from a licensed broadcaster. This means, take a GOOD digital tuner and scan the band at 6am, 12pm, 6pm, 12am. Note all of the stations you find and their strength. You ideally wish to find a frequency with no immediate adjacent channels in use (of course this is extremely tough). In other words, if you find that 630kHz is not in use, you would want to also determine if 610, 620, 640, 650kHz were either vacant or distant. In the ideal world that would be the case. Since this isn't an ideal world, you try to find the best 3–5 station gap you can. Try to start at 530kHz (usable for carrier current and highway advisory but not licensed commercial broadcasters) and work your way up the dial. Good Luck!

Anyway, with that basic info out of the way—send me your snail mail info and I will send a pile of info for you (same for anyone else needing an update!).

Thanks. John Devecka, LPB, Inc.

Goodday. I would like to post a question concerning low wattage AM broadcasting. Our station is looking into the addition of a training wing that will also serve in better serving the community with more news and sports coverage.

What do we need to consider when looking into this proposal? We plan to cover the campus of the University of Iowa which, for our purposes, is about a 5 mile radius. How do we select frequency and call letters? If anyone has been in a similar situation, please respond.

thanks, operations director

Received: from gatekeeper.fcc.gov by gwuvm.gwu.edu;
Thu, 08:55:55 EDT #40
From: Allen Myers, FCCamyers@fcc.gov
Subject: AM BROADCAST-Reply

The service which you want to provide requires an FCC license, which means that you must conduct a frequency search to find an available frequency and file an application with the Commission for an AM broadcasting station. The only low-wattage AM broadcasting system which we allow without a license is AM Carrier Current. This operates at 100 mW power and will provide service within a building and about 200 feet from it.

—Allen Myers, amyers@fcc.gov,

# Appendix C ▶ ▶

# WMHW-FM Handbook of Policies and Procedures

Jerome D. Henderson

## General Policy and Objectives

*WMHW-FM Radio Station is owned by Central Michigan University, Mt. Pleasant, Michigan and is licensed by the Federal Communications Commission.*

In accepting its license, the station accepts as its primary purpose and function, as stipulated by FCC Rules and Regulations, the responsibility to "serve in the public interest, convenience, and necessity." The licensee is obligated to make a continuing, active search to ascertain the needs of the community served and to fulfill those needs to the best of its ability.

Therefore, one primary purpose of WMHW-FM is to serve in the public interest, convenience, and necessity within its assigned listening area.

WMHW-FM is licensed to an educational institution as a non-commercial radio station. The station is under the direct supervision of the Department of Broadcast and Cinematic Arts and operates as a primary supervised radio broadcast laboratory for students of that department.

Therefore, by nature of its ownership and curricular affiliation, a second primary purpose of WMHW-FM is to provide a co-curricular radio broadcast training experience for talented students interested in broadcasting as a career or exploratory experience.

WMHW-FM will provide opportunities for research and innovation in creative programming utilizing the full resources of the department and the curriculum. In addition, because of its non-commercial licensure, WMHW-FM will provide, where feasible and appropriate, an alternative programming service to Isabella County by broadcasting features not otherwise available to area radio listeners on other radio stations.

WMHW-FM will at all times comply with the spirit and letter of the Fairness Doctrine. Recognizing its position as a non-commercial licensee of a state educational institution, WMHW-FM will not editorialize.

By virtue of its ownership, WMHW-FM is a public image source for Central Michigan University. The station will constantly endeavor to enhance the image and visibility of the university and the Department of Broadcast and Cinematic Arts through the production and broadcast of high quality local programming. The station is not a primary disseminator of information from the university to the public since that is a function of the Media Relations Office of the university. All requests from sources outside the university for "direct feeds" of information concerning university events or activities should be referred to the Media Relations Office.

Opportunity to be involved in the activities of WMHW-FM shall not be denied because of religion, race, color, national origin, age, sex, marital status, veteran status, handicap, and/or sexual orientation.

WMHW-FM, as a responsible broadcast licensee, shall at all times retain the right of final judgment over programming decisions.

II.   Equipment Rooms Policies

The following are rewritten from all-inclusive BCA Department policies to reflect specifically activities at and concerning WMHW-FM:

1.   No eating, drinking or smoking is permitted in any BCA equipment rooms, including WMHW-FM Master Control Room, Production Control Room, News Room, and 8-Track Studio. A first violation will result in that student's suspension from the use of all BCA equipment for a two-week period. A second violation will result in an immediate one-semester suspension from the use of all BCA equipment.

2.   Changing of connections to any piece of radio studio/control room equipment will be done only by the engineers. The equipment has been installed to achieve most of the things necessary to basic radio production and broadcast. Further flexibility is available through the use of the audio patch panels. Therefore, the changing of any connections must be done by someone who understands the system and the effects which might result from connections changes. Switching of cable connections will result in the immediate suspension of the student's use of all equipment for a period of two weeks.

3.   Each school day, the Radio Operations Manager will do an equipment check-out in order to ascertain the operating status of all equipment in WMHW-FM radio facilities. Any malfunctioning equipment will be identified with an "Out of Order" tag and no attempts are to be made to use such equipment until it has been repaired.

4.   Any student taking out WMHW-FM equipment or using the WMHW-FM facilities will sign a form stating that he/she is responsible for its care. If a piece of equipment ceases to work properly while in use, or is found out-of-order when any use is intended, it is the student's responsibility to inform the Radio Operations Manager of such in writing of:

(a)  Which specific piece of equipment is out of order,

(b)  Date and time of the occurrence, and

(c)  Conditions under which it went out of order, if known.

For any equipment found to have been misused or abused, the student responsible will (1) be liable to suspension from all BCA equipment use for a two-week period and/or (2) be charged for the repairs to the equipment. The BCA department has the authority to have the university hold grades of students until all outstanding bills are paid.

5.   Unauthorized use or abuse of equipment/facilities will result in a student's immediate suspension from all BCA equipment use for a period of one month.

III. Station Organization

A.   Central Michigan University, as governed in descending order by the Board of Trustees, President, Provost, Dean of the College of Arts and Sciences, and Chairperson of the Department of Broadcast and Cinematic Arts, authorizes overall station policy, operating function, and financial activity.

B.   The General Manager of WMHW-FM is the Chairperson of the Department of Broadcast and Cinematic Arts and is responsible for all activities of the station.

C.   The Radio Operations Manager is appointed by the Chairperson of the Department of Broadcast and Cinematic Arts and is responsible to the department for day-to-day operations of the station as well as any activities representing the station.

D.   The Executive Council of WMHW-FM consists of the General Manager, the Radio Operations Manager, and those students appointed to Executive Staff positions by the

Department of Broadcast and Cinematic Arts. The Executive Council includes:

1.  General Manager*

2.  Radio Operations Manager*

3.  Station Manager

4.  Program Director

5.  Business/Grants Manager

6.  News Director

7.  Sports Director

8.  Promotions Director

9.  Production Director

10. Music Director

11. Traffic Director

12. Marketing Research Director

13. Senior Producer of Elections (Fall semester, election years only)

*Denotes University appointed faculty/staff

The structure of the Executive Council of WMHW-FM embraces the department concept of management. It is highly functional in nature. The members of the Executive Council are equal in voting stature.

Chairperson of the Executive Council is the Station Manager. In the Station Manager's absence, the Program Director assumes leadership.

The Executive Council sets both immediate and long-range goals for the station and works to assure that goals are achieved. In doing this, among other activities, the Executive Council makes major programming decisions on a long-range basis, coordinates station activities, and reviews and evaluates the progress of station personnel. The Executive Council also approves appointment to any supervisory management positions on the station staff.

E.  The Procedure for Appointment to Executive Council involves the incumbent Executive Council, the Radio Operations Manager and the faculty/staff of the Department of Broadcast and Cinematic Arts.

Executive Council positions normally are selected near the end of the Spring Term for the following regular academic year. In addition, selected positions may be filled as deemed appropriate by the Department for the summer months. Further vacancies may occur from time-to- time by resignation or other causes.

When a position is to be filled on the Executive Council, notice of that pending or current vacancy is posted on the Department and radio station bulletin boards. Any student who wishes to be considered may apply for the position(s) so posted by completing the Executive Staff Application Form and submitting it to the Radio Operations Manager along with any other supporting documents on or before the submission deadline posted on the vacancy notice.

At the next meeting of the Executive Council, interview times will be established for each applicant for each vacancy posted. All candidates for a position will be interviewed, after which the Executive Council will deliberate and recommend the name of the most acceptable candidate to the Radio Operations Manager by secret ballot. The Radio Operations Manager reviews the Executive Council recommendation(s), attaches his/her own recommendation(s) and forwards both sets of recommendations to the Department of Broadcast and Cinematic Arts.

The Department, after deliberation, makes the final decision on the appointment of the person to fill each vacancy.

The Executive Council, the Radio Operations Manager, and the Department all have the option of voting to reopen any position for which a suitable applicant has not been determined. In such case, the notice of reopening of the position vacancy is posted and the above procedure is repeated. Persons who originally applied for a position that has been reopened may reapply for that position.

## Notes

Executive Council positions are typed in bold face.

Senior Producer of Elections is appointed only in Fall Semester, election years.

Supervisory Staff positions, except for PSA Director, are not specified in this chart.

IV. Duties and Responsibilities

A. General Statement—All student Executive Staff personnel are responsible for fulfilling the following roles:

1) Conduct ongoing performance evaluations of all persons working within their departments, including a written evaluation of each individual at the end of the semester to be submitted to the Radio Operations Manager for inclusion in that individual's personnel file at the station.

2) Monitor station activities, especially those pertaining most directly to their departments, for the purpose of continually improving the station's on-air sound and operations as well as the public image of the station presented in any situation.

3) Attend all Executive Council meetings and general station staff meetings, prepared for the deliberations that will take place at those meetings.

4) Prepare a budget request in a timely manner when called for at Executive Council; and monitor and control all budgetary activities within their own departments so as to keep within the budget authorized.

5) Continually work with all personnel within their departments in a manner that furthers each one's broadcast education and assures well-trained applicants for Executive Staff positions in future semesters.

6) Strive to improve their own professional demeanor and performance to be able to present the best possible role model for those in their respective departments.

7) Represent the station at all station functions, insofar as possible.

In addition, each of the Executive Staff positions includes specific duties and responsibilities, many of which are listed on the following pages.

B. Station Manager—The Station Manager supervises the daily activities of the station by guiding the departments to set and accomplish goals. He or she is responsible for all additional duties delegated to him/her by the Radio Operations Manager or General Manager concerning any aspect of the operations of the station. The Station Manager chairs all

Executive Council and general staff meetings and is responsible for the recording and appropriate dissemination of minutes of those meetings. Coordinating the activities of the various departments, the Station Manager works closely in association with department directors while overseeing all aspects of station functions. He/she is a liaison between the BCA Department and station personnel. When possible, the Station Manager along with other department heads should represent the station at all station functions.

C.  Program Director—The Program Director is responsible for all on-the-air program functions of the station with the exception of those responsibilities and operations delegated to other department heads (e. g.: News Director, Sports Director, etc.). All scheduling of programs as authorized by the Executive Council, coordinating with the News and Sports Directors, falls within this function. The Program Director is also responsible for the selection and appropriate training of all board operators and for regular ongoing monitoring of their activities through air-checks, training and critique sessions. It is his/her duty to coordinate the activities of all programming personnel in a manner as to assure that all air shifts are filled and the program format is followed. In addition, the Program Director is responsible for assuring that all programmers (board operators) have their appropriate FCC permits/licenses posted in the station Master Control Room (MCR). Under individual circumstances, the Program Director may have new duties and responsibilities assigned to him/her on a temporary or permanent basis after consultation with the Executive Council.

D.  Music Director—The Music Director is responsible for the music playlist used within the program format and for transmitting any and all appropriate information regarding the playlist to on-air personnel, the Program Director, record companies, and the general public as deemed advisable by the Executive Council. The Music Director is directly responsible to the Program Director; and also is responsible for continual updating of the playlist, the card and record filing and marking systems, and general security of the music records and CDs of the station. when the Music Director and Program Director are unable to agree on the appropriate classification for, or whether a particular piece of music should be on the station playlist, the Program Director's authority normally supercedes.

E.  Production Director— The Production Director is responsible for quality control of all pre- produced materials used on-the-air and also is responsible for scheduling the use of the Production facilities and for maintenance of the Sound Effects and Production Music libraries of the station. He/she is directly responsible to the Program Director and works cooperatively with other members of the Executive council to assist in the development of new program materials for station use. He/she also is responsible for overseeing day-to-day activities of the production staff and to provide appropriate training and critique for them. The Production Director has the authority to reject any and all production materials not deemed of high enough quality for airplay on the station. He/she is to maintain a general surveillance of equipment and production materials for purposes of control and supply. In addition, the Production Director should be familiar with all the components of production and should strive for creativity in production.

F.    News Director—The News Director fulfills the programming function in news and public affairs oversees the operation of the news department and other public affairs programming, and coordinates directly with the Program Director in determining scheduling of newscasts and other news and public affairs programming. He/she is responsible for guiding the training of all personnel within the department, auditioning and selecting newscasters, improving newswriting skills, developing format and requirements of newscasts, and regular monitoring of news activities through air checks, training and critique sessions. In addition, the News Director is responsible for continually developing new contacts for news information, maintaining local (and regional, where feasible) news-beat assignments for personnel, providing equipment for newscasters to enhance their news reporting, and developing ideas for covering areas of news interest. In addition, the News Director is responsible for keeping accurate updated files ,concerning issues of local interest or controversy and any news or public affairs programming that has been aired in relation to those issues. The New Director is to monitor the AP teletype regularly to assure proper operation, sufficient paper supply and good quality ribbons.

G.    Sports Director—The Sports Director is responsible for all programming related to sports broadcast on the station. He/she must audition and appoint all personnel who prepare and/or broadcast regular sports reports as well as play-by-play sportscasts. The Sports Director is generally responsible for any sports actualities used in broadcasts. He/she is the senior producer for all live sports broadcasts and therefore is responsible for all activities related to preparation, broadcast, and follow-up of those broadcasts, including (but not necessarily limited to) preparation of and control over budgets for each event, scheduling of needed broadcast lines, and transportation and room and board arrangements for sports personnel involved. The Sports Director is also responsible for appropriate training and critique of all sports department personnel.

H.  Business/Grants Manager—The Business/Grants Manager of the station has two primary responsibilities:

1)  To seek, on a continuing basis, new and repeat funding to support station activities through business, professional, and personal grants or donations to the station and the Department of Broadcast and Cinematic Arts; and

2)  To authorize budgets for the departments within the station in accordance with the guidelines established by the Executive Council, oversee all activities within each of the various budgets, and guide the department heads of the station in adequate preparation and appropriate use of the—budgets allotted to their departments.

In accomplishing these, the Business/Grants Manager works directly with the Radio Operations Manager and any other person assigned by the BCA Department to oversee accounts in order to assure full compliance with university guidelines and requirements for such activities. It is his/her responsibility to assure that all personnel on the Business/Grants staff are fully trained in all procedures required by the university and the BCA Department, to train, coach and critique all station staff members involved in grants activities in the best procedures for approaching constituencies in order to secure grants, and to coordinate all activities in this area.

It is further the responsibility of the Business/Grants Manager to secure budget requests from all station department heads and to prepare the overall station budget for presentation to the Executive Council at the start of each semester. Income generated from conducting dances and other professional broadcast activities of station is entered into the station general account and is controlled by the Business/Grants Manager upon receipt at the station.

I.  Promotion Director—The Promotion Director is responsible for all promotional and public relations activities of the station, including both on-air and in-station promotions of station events and activities as well as coordinating and maintaining an appropriate station image in non-broadcast activities. The role of this position is two-fold, both external to the publics the station serves and internal to communicate station activities to the entire staff. The Promotion Director is the key liaison in all matters pertaining to the presentation of the station image. He/she is responsible for regular contact with the university's Office of Public Relations, preparation and dissemination of press releases (in coordination with the Public Relations Office where deemed appropriate), coordinating all materials for bulletin boards and/or distribution to the general public, and assisting the General Manager, Radio Operations Manager, Station Manager, and other station staff personnel in preparation of materials and presentation of the station to external groups or individuals. He/she also is responsible for training and coordinating a staff to help fulfill these responsibilities.

J.  Traffic Director—The Traffic Director is responsible for the orderly preparation and use of the Program and operating (technical) Logs of the station. He/she assures that all logs are properly completed, any needed corrections are made, and that all regulations of the Federal Communications Commission (FCC) pertaining to station logs are followed and is responsible for coordinating all grants/donor announcements scheduling according to FCC Rules and Regulations. The Traffic Director also coordinates the broadcast of Public Service Announcements through a PSA Director responsible to him/her.

K.  Marketing Research Director—The Marketing Research-Director is responsible for coordinating, conducting, and evaluating all research appropriate for the station. It is his/her responsibility to assure that measurement instruments and research techniques used are valid and reliable and that results are reported to the Executive Council within the reasonable time requested. Expected research includes, but is not limited to: various listener surveys, community needs ("issues") surveys, and occasional focus group studies on a narrowly defined topic. The Marketing Research Director is directly responsible to the Station Manager and works closely with the Executive Council in defining criteria and parameters for surveys. Regular progress reports on research being contemplated and conducted for the station are presented to the Executive Council.

L.  Senior Producer of Elections—The Senior Producer of Elections (appointed by the BCA Department for the Fall Semester of election years) is responsible for coordinating all activities of the station relating to the upcoming November election. This includes maintaining accurate files pertaining to all local, major State and National election issues and candidates, overseeing research, writing, and production for airing of all election-related programming, appropriate response to all inquiries to the station concerning election matters, and preparation and production of Election Night programming. It is his/her responsibility to select, assign election staff personnel to these tasks. He/she is directly responsible to the News Director and according to guidelines established by the Executive Council, and is responsible for the election budget.

M.  Other Appointed Station Personnel (Supervisory Staff) Positions that are important to the operation of the station, though not Executive Council level, may be designated by the various department heads when approved by the Executive Council. Examples of these include Assistant News Director, Assistant Program Director, etc. Two such positions, because of their importance, are detailed below.

When the Executive Council approves a Supervisory Staff position, the appropriate department head advertises the position, accepts applications, interviews applicants, and presents his/her recommendation(s) to the Executive Council which makes the appointment of the individual to the specified position.

Executive Secretary—The Executive Secretary attends all meetings of the Executive Council and general station staff to record the minutes of the meetings. These minutes are posted and otherwise appropriately distributed in a timely manner and are filed for future reference. The Executive Secretary is appointed upon the recommendation of and is responsible to the Station Manager.

PSA Director—The PSA Director is responsible for coordinating all requests from agencies and persons outside the station to have Public Service Announcements aired on the station. He/she reviews all such request, selects such announcements for broadcast, and writes and/or produces the materials necessary to assure accurate airing of them. The PSA Director is directly responsible to the Traffic

Director and submits all PSA materials prepared to the Traffic Director for appropriate scheduling in the Program Log.

V. Regulations for Working at WMHW-FN

A. All students working at the station must have and maintain a minimum overall Grade Point Average (GPA) of 2.25.
All students applying for and/or holding Executive Staff positions must have and maintain a minimum overall GPA of 2.50.

B. Student Executive Staff Personnel will volunteer, or will register for Practicum credit (BCA 329), for services rendered in those positions. It is the responsibility of the Radio Operations Manager to determine the number of hours to be earned by Executive Staff personnel and to act on requests for Practicum registration.

C. Regular staff members will volunteer, or will do Practicum work. It is the responsibility of the Radio Operation Manager, with the advice of the Executive Council, to oversee workloads for all station personnel, including whether such workloads may be counted toward Practicum credit (BCA 329).

D. Volunteer students
Any university student who wishes to volunteer his/her services at WMHW-FM for the purpose of gaining experience in the field and/or as an avocational interest may make application to do so. The nature of his/her service is governed by the student's own interest and aptitude, and will be utilized as determined by the appropriate Executive Staff member and the Station Manager. Each such student will sign the standard WMHW-FM Personnel form and abide by station and BCA Department policies.

E. Practicum Students (BCA 329)
Students who wish to register for Practicum credit (BCA 329) must receive approval from the Radio Operations Manager prior to registration for the course. In addition, Practicum students must sign the standard WMHW-FM Personnel form and abide by station and BCA Department policies.
Practicum students are under the guidance of the Radio Operations Manager, who approves where the students work at the station, with the advice of the appropriate Executive Staff member(s). Such students will abide by the BCA Practicum Policies, which apply to several practicum-granting activities. The grade for the Practicum student shall be determined by the Radio Operations Manager after

consultation with the appropriate Executive Staff member(s).

F.    Practicum Prerequisites
Prerequisites for Practicum credit for certain activities at WMHW-FM, established by the BCA Department, are listed on the following page. Any student who registers for Practicum credit without prior approval of the Radio Operations Manager and who does not have the necessary prerequisites for the position he/she wishes to fill may be required to withdraw from BCA 329.


STUDENT STAFF Opportunities, WMHW-FM

Students may participate in activities of WMHW-FM with the approval of the station Radio Operations Manager, throughout their academic careers at CMU. Students desiring practicum credit for station work normally should have completed the prerequisites listed below. Completion of courses in parentheses, while not required, may give an applicant preferred status.

| | |
|---|---|
| Station Manager | 222, 313, 510, (516), (519), and two semester Executive Staff experience |
| Program Director | 222, 313, 317, (510), (516), (519) and one semester Executive Staff experience |
| Grants/Business Mgr. | 511, 516, or relevant business course |
| News Director | 222, 317, 318, (518) and one semester experience in news |
| Sports Director | 222, 317, 318, (518) and one semester experience in sports or news |
| Music Director | 222, 313, (421), (519) |
| Production Director | 222, 317, 421 |
| Promotions Director | 311 512 |
| Marketing Research Dir. | MKT 350, (MKT 450), (BCA 516) and one semester experience in Marketing Research |
| Traffic Director | FCC Operators Permit, (519) |
| News Reporter | 317, 318 |
| Air Personality | 222, 317 and FCC Operators Permit |
| Sports Reporter | 317, 318 |
| PSA Director | 222, 311, 317 |
| Producer, Special Progs. | 222, 311, 313, 421 |

All Executive Staff positions (indicated in bold type) generally should be filled with students enrolled for Practicum credit (BCA 329).

VI.  Conduct and Responsibilities of Station Personnel

WMHW-FM will adhere to and be governed by all regulations set forth in this document and the University's "Policy and Procedures Relating to Students Rights and Responsibilities".

A. No Longer Applicable

B. Broadcasting Responsibility

The Program and Operating Logs are university documents which show compliance with certain FCC Rules and Regulations and must be maintained accurately. Therefore, the person whose signature appears on these Logs for a certain period of time is fully responsible for the materials broadcast during that time and for proper operation of the transmitter. Any and all irregularities must be recorded on the appropriate Log and brought to the attention of the Program Director. In like manner, staff personnel in the radio station are responsible for actions that take place in the station. Each must check to make certain that any visitors or telephone callers are well treated and that strangers are offered assistance and have specific business in the station. Each staff member is responsible for the security of the entire station, including the record library holdings. If any staff member has questions, he/she should contact the Program Director or other Executive Council member.

C. Station Use After Hours

Those staff members wishing to use station facilities after 11 p.m. should see either the Station Manager or the appropriate department head for permission.

D. Noise

Noise levels are to be kept at a minimum everywhere in the station. Faculty, staff, and graduate assistants are active in non-station pursuits in other offices in the station wing.

E. Studio Monitors

Studio monitors are to be kept at a reasonable level to avoid damage to the speakers and avoid interference with other activities at the station or in the station wing.

F. Discipline For Station Staff

Staff members will be suspended or dismissed from all duties and office for:

1. Blatant disregard for Log procedures.
2. Unexcused failure to report for assigned duties (board shifts, news and sports assignments, staff meetings, production, etc.)
3. Removal, without permission, of any station property, including records. (Public Safety will be notified in the case of all apparent thefts. Disciplinary action will be

       taken either through the Office of Student Life or criminal prosecution.)

4. Profane, obscene, or indecent language on the air.
5. Playing of any record or tape which is blatantly obscene, profane, or indecent or which recommends use of illegal drugs.
6. Inappropriate conduct, while representing the station, which would be detrimental to the image of the station, the BCA Department or the University.
7. Any action which violates FCC Rules and Regulations.
8. Any other specific action which results in censure by unanimous vote of the Executive Council.
9. Possession or use in the station of alcoholic beverages, illegal drugs, dangerous chemicals, firearms or explosives, as well as found being impaired by alcoholic beverages or illegal drugs.
10. Tampering with (or "adjusting" or "fixing") any station equipment.
11. Failure to attend mandatory meetings called by the Station Manager or by any other Executive Staff member for his/her specific department. Absences should be cleared in advance. Unexcused absence from a mandatory meeting will result in suspension from activities in the specific department involved for a period of two (2) weeks. Repeated unexcused absences from mandatory meetings will result in dismissal from WMHW-FM.

G. In each case of violation of station policies and procedures, some of which are listed in "F" above, the Executive Council will determine if the incident is severe enough for suspension or dismissal and will inform the involved staff member(s) of its decision. All suspended or dismissed staff members will be notified in writing, and a copy of that notice shall be placed in the student's personnel file at the station.

H. APPEALS of the Executive Council decision for suspension or dismissal may be presented in writing to the Radio Operations Manager within ten (10) days of the decision by the person suspended or dismissed. The Radio Operations Manager will review all facts and assertions presented both to the Executive Council at the time of its deliberation as well as in support of the appeal and will pass his/her recommendations on to the Station General Manager within five working days. The General Manager will make a final decision on the appeal.

I. Logging Violations

Programmers will receive written notice from the Traffic Director of all Program and/or Operating Log violations. For those who do not correct their log violations within a week, the following actions will automatically be taken:

First Offense: Suspension of all on-air duties for a period of four (4) weeks of time the student would normally be in school.

Second Offense: Suspension of all on-air duties for period of 16 weeks of time the student would normally be in school.

Third Offense: Dismissal from the station for the remainder of the student's academic career at CMU.

Other: Extraordinary situations, as determined by the Executive Council, will be dealt with separately.

All Operating Log violations are subject to entry of such in the student's personnel file at the station.

J.    Station Personnel File

A file will be maintained in the office of the Radio Operations Manager on each student involved in the station. This file will be accessible upon request as follows: (a) to the student, (b) to the Executive Council during deliberation of personnel matters involving the student, (c) to the Station Manager, (d) to the Radio Operations Manager, and (e) to the General Manager.

The personnel file of a student may be used when he/she applies for a station Executive or Supervisory staff position, requests a letter of recommendation, applies for a Special Talent Scholarship, or is under consideration for suspension or dismissal from the station.

In this file will be kept all records of activities by that person at WMHW-FM, including statements and evaluations concerning work performance and any disciplinary actions taken, applications for staff positions and whether or not awarded, statements of achievements of note while working for the station, awards presented by the station or the BCA Department, and other materials deemed pertinent.

Disposition of File: The "Active" file will pertain to all personnel currently on the station staff or who were on staff during the prior regular semester. When the person leaves the station, his/her records will be retained in an "inactive" file for a period of one calendar year for those not yet graduated. For those who graduate, pertinent materials from the files of BCA Majors will be transferred to the appropriate file in the BCA Department main office. Files of non-majors will be retained for a period of two (2) years in the office of the Radio Operations Manager, in the event of requests for letters of recommendation. "Inactive" and "Graduated" files kept in the Office of the Radio Operations Manager will be discarded at the end of the stated retention period.

Appeal Procedure: If an individual believes that materials in his/her file are misleading or inaccurate, an appeal may be made to have such materials corrected or removed from the file. The appeal will follow procedures listed below:

(a) The individual must submit a letter of appeal to—the Radio Operations Manager detailing in specific which materials are felt to be false, misleading, or inaccurate and also what he/she believes would be appropriate. An exact copy of the letter of appeal is to be filed with the station's General—Manager at the same time.

(b) Within a reasonable time, but not more than ten (10) working days, the Radio Operations Manager will arrange a meeting with the individual fill the appeal at which time the individual may present any evidence to support the appeal claim(s). A written record of this meeting will be made, with copies to the individual, the Radio Operations Manager and the General Manager.

(c) The Radio Operations Manager will then present recommendation on the appeal in a timely manner the General Manager who will make a final decision and present that decision in writing to the individual and the Radio Operations Manager.

(d) During the time that an appeals decision is pending, any materials in question will not be used in matters concerning the individual.

(e) The final decision by the General Manager is binding; and at the time of the decision, appropriate action will be taken as regards the individual's personnel file at WMHW-FM.

VII. WMHW-FM Departmental Policies

A. Program Department

Programmers are responsible for any and all materials aired when they are signed on the Program Log. Any abusive, profane or obscene language either by the programmer or within his/her choice of music will be dealt with by the Executive Council. Care is to be exercised to prevent the use of material suggesting drug orientation. Any music unfamiliar to the programmer is to be previewed before airplay.

Unexplained absences from program assignments are not acceptable; and removal from program responsibilities may be invoked by the Program Director if any such absences occur.

Only those persons directly involved in a program in progress or about to be aired or taped are to be in the MCR, The programmer on the air has the prerogative to request the departure from the MCR of any person or persons disrupting program operation, with the exception of working engineering personnel. The number of people in the MCR should be kept to a minimum to avoid interference with programming.

All complaints received by any person concerning programming on the station should be carefully noted and referred to the Program Director and Station Manager. Also, they should be dealt with politely. Any programmer who must be absent from his/her assigned program must notify the Program Director as to when he/she will be absent and what competent substitute has been found for that date.

Programmers should not deviate from their specific show format.

Test of the Emergency Broadcasting System (EBS) incoming on the AP teletype are to be attached to the Operating Log and noted in the appropriate space on that Log. EBS tests incoming on the monitor in the MCR are also to be noted appropriately on the Operating Log. EBS tests listed in the Program Log are to be run as scheduled and noted appropriately on the Operating Log.

B.    News and Sports Departments

News, sports and other items coming off the AP teletype are to be "pegged" or otherwise filed in their respective categories as soon as they are taken off the teletype.

Newscasters and sportscasters who are to be announcing for a regularly scheduled news or sportscast should be at the station preparing for it at least one hour before the time of the cast.

Public Affairs or pre-produced news tapes and cartridge tapes (carts) are the property of the news department. Sports tapes and carts are the property of the sports department. Unless otherwise labeled, other station personnel may not touch.

If a substitute is needed for any news or sportscast, notify the appropriate department head (News or Sports Director) immediately and follow his/her instructions.

All problems with the AP teletype machine are to be reported to the News Director. It is his/her responsibility to ensure that problems are corrected as soon as possible.

C.  Music Department

Records or CDs may be taken from the radio station only when authorized by the Executive Council or the Radio Operations Manager.

The programmer on the air always has priority to the audition turntable in the Music Library if needed for a quick review. Programmers previewing music for an upcoming show have next priority.

Previewing of records is to be done in the record library, not in the MCR or Production Control Room.

D.  Sign-On Procedures—See Station Day Book

E.  Sign-Off Procedures—See Station Day Book

F.  Program And Operating Log Procedures

When a programmer (board operator) comes on duty, he/she is to enter time and signature on the page of the Program Log where his/her responsibilities begin. Each subsequent page of the Program Log upon which he/she makes entries must also be signed. Signature and time off must be entered on the last page for which he/she is responsible.

Programs more than one minute in duration require an end time as well as a starting time to be logged.

The Operating Log must also be signed, indicating time on and time off. Regular operating meter readings must be taken as required.

G.  Log Corrections

Do not obliterate or erase anything. Simply draw a line through the incorrect entry and write the correction in the appropriate place. Corrections made before or at the time of broadcast must be initialed and dated by the person making the correction. Corrections made after the operator has signed off the log must also include the reason the change was made.

VIII.  Non-Broadcast Events Using WMHW-FM Equipment

From time-to-time, WMHW-FM or its personnel are asked to DJ dances and other non-broadcast events using BCA equipment (KG: "Dance Board", amplifiers, speakers, CDs, records, etc.) used by WMHW-FM. The following policies and procedures apply for this purpose:

1.  Use of any and all WMHW-FM equipment for dances and other events not directly related to on-air operations must be coordinated through the Promotion Director of the station.

2.  If the station is receiving financial remuneration ("pay") for doing an event, the amount of renumeration is subject to approval of the Promotion Director

and the station Executive Council and may be used to determine whether or not said event will be done using station equipment.

3.  The Executive Council of the station is the sole determining body as to whether or not personnel performing in behalf of the station (DJs) or using station equipment are to be paid for doing an event.

4.  In general, DJs performing such events will receive pay only when the station receives financial remuneration. Said pay for any DJ, or all DJs combined, for any single event shall not exceed one-third (1/3) of an after-expenses remuneration received by the station. An example of such "after expenses" would be for equipment that has been rented by the station for the event.

5.  Payment to DJs for such performances shall be made following normal and accepted university policies and procedures (e.g.: through the university's payroll office, using appropriate vouchers, etc.).

6.  Donation of a payable fee by DJ(s) to the station shall constitute a recognizable donation to the university and will be handled as such through regular channels (e.g.: Development Office).

    Variances from the "normal" expected procedures (as stated above) shall be permitted only by approval of the Radio Operations Manager and/or majority vote of the station Executive Council unless required by superior authority of BCA Department or university policy or procedure.

VIII.  Modifications to this Document

Proposed changes to this document initiated within the station will be recommended by the Executive Council to the Broadcast and Cinematic Arts Department for consideration. Upon approval, they become part of the Policies and Procedures of WMHW-FM.

# Appendix D ▶ ▶

## Getting On the Air at WGFR

Most new broadcast students at Adirondack Community College want to go "on air" at the campus radio station, WGFR, promptly. A Permit or "license" is no longer required. When you begin your classes you will receive training in each essential step of broadcast station operation. Meanwhile you must know professional broadcast law. Here are FCC laws and rules:

WGFR is licensed to Adirondack Community College. The station belongs to the College, not to the students.

All program material must meet standards of propriety and taste appropriate for a radio broadcast station licensed to a public community college.

Copyrighted material may not be broadcast, except music for which WGFR pays royalties. Licensee policy does not allow broadcast of comedy recordings.

Broadcast the station identification, the "ID," at sign on time, sign off time, and on the hour, according to FCC Rules 73.1201. Say "WGFR, Glens Falls." Do not say "WGFR-FM." Though WGFR is an FM station, "FM" is not part of the legal call letters. Insert nothing between "WGFR" and "Glens Falls."

No business, product, or service may be mentioned directly or indirectly (with the limited exceptions of formal underwriting and similar arrangements made by management). Never say something such as, "This recording goes out to all the people at Joe's Diner." Don't even say something such as "The fast food place with the golden arches." Also, never mention parties and other events on the air.

The Emergency Alert System, EAS, must be monitored correctly. Instructions are posted at the station and all operators are trained how to receive and log any Test Alert received and where to find the Checklist in an actual emergency.

Answer the telephone in a friendly manner. If a caller is a nuisance or abusive, politely and courteously hang up. No business tolerates an employee who is rude on the telephone, and the same is true at

WGFR. Keep the phone free for requests and avoid extended conversations with friends. WGFR serves the public.

Control room and studio doors remain closed. Smoking is not allowed.

Students have no authority to perform technical work, connect or disconnect hardware, or interface their own equipment with WGFR facilities.

Never mention parties or private functions and events on the air.

Payola is accepting or agreeing to accept anything of value in return for broadcasting records or promoting products or events on the air without disclosing that payment has been made. Section 507 of the Communications Act specifies a $10,000 fine and up to a year in prison.

Federal statutes provide fine and imprisonment for a person who "has failed to carry out the lawful order . . . of the person in charge," "has willfully damaged or has permitted radio apparatus or installations to be damaged," or "has transmitted superfluous communications or . . . profane or obscene words, language, or meaning."

Music and programming specifically disallowed includes but is not limited to:

1.  Racist, religious, or sexist slurs.

2.  Condoning illegal practices.

3.  Condoning use of illegal drugs.

4.  Glorifying tobacco and alcohol. 5. Obscene and
    indecent material.

Broadcasting of obscene or indecent material is prohibited by the Communications Act and by Section 1464 of the U.S. Criminal Code, with a fine of up to $10,000 and/or imprisonment up to two years.

Indecency is defined as "language or material that depicts or describes, in terms patently offensive as measured by contemporary community standards for the broadcast medium, sexual or excretory activities or organs. " The Federal Communications Commission itself decides what is 'patently offensive," and it defines "contemporary community standards."

1.  The F.C.C. doesn't consider the popularity of a program, recording, etc., in deciding whether material is indecent.

2.  Airing such material based on past F.C.C. inaction in regard to similar material is no defense.

3.  Innuendo and double-entendre do not save a statement from indecency when "clearly capable of specific meaning." The National Association of Broadcasters advises that the safest way to avoid difficulty is"to stay far wide of any broadcasting of any material as to which the licensee harbors any doubts at all," and this is the policy for WGFR, incumbent upon the faculty station manager, the student Program Director and Music Director, and all students performing on the air.

*In denying license renewal to WXPN, University of Pennsylvania the F.C.C. said that "licensee control over the operations and management of broadcast facilities has been central to the proper functioning of the regulatory scheme mandated by Congress . . . licensees are permitted to delegate . . . day-to-day supervision . . .but . . . a licensee, educational or otherwise, may not delegate authority over a broadcast facility . . ." Report No. 14507, Docket 20677

92.7 Mhz Glen Falls, NY
Program and Operating Log

Day
Date        199X

EBS Test Alert received from WMJR at
Transmitter On time: *8:07AM*
Transmitter On time:
Transmitter On time:

(time). Signature
Signature    *Leslie Doe*
Signature
Signature

Operators signed on and off below are on transmitter duty and program log duty. Each
operator signs on when going on duty, and again, separately, when going off duty, attesting
that this log accurately represents actual programming and transmitter operation during
his/her period of duty.

*Leslie Doe*        time on *8:07*    Leslie Doe    time off *11:00A*

| Sche-duled | Actual Time | Program Material |
|---|---|---|
| 8:00A | *8:07* | SIGN ON CART WITH LEGAL ID |
|  | *8:10* | PSA *Red Cross Blood Drive* |
|  |  | PSA |
|  |  | PSA |
|  |  | PSA |
| 9:00 |  | CORRECT TIME AND LEGAL ID |
|  |  | PSA |
|  |  | PSA |
|  |  | PSA |
|  |  | PSA |
| 10:45P |  | SIGN OFF CART WITH LEGAL ID |
|  |  | TRANSMITTER OFF (log on other side too) |

Here's part of the cover page of the WGFR log. Note the transmitter on and off time, and where each operator (DJ) signs on and off. Also notice where you fill in and sign receiving an EBS Test Alert.

Follow the log (see sample to left) and do what it says. Be sure to send the legal ID on at the top of the hour, and be sure it is legal ("WGFR, Glens Falls")Fill in the "actual time" column for every line.

Make all entries in ink. (no pencil or crayon). Make no entries except those required (no requests or phone numbers).

Do entries needing corrections this way: cross out the incorrect entry with a single horizontal line through the entry (like this), and then write the proper entry above it and place your initials next to the correction. Never erase or obliterate. The log is a legal document!

If there is no log available when you start your shift, then keep a written record of your activities and bring that report to the traffic manager as soon as possible.

Make all entries absolutely correct. Don't change times and don't sign your name for any else.

## Station Identification

Radio announcers and television control operators broadcast station identification. When? According to FCC. Rule 73. 1201, "At the beginning and ending of each time of operation, and . . . hourly, as close to the hour as feasible. at a natural break in program offerings."

In other words, identify your station at sign on and at sign off, and on the hour. If you are broadcasting music, time your recording to allow reading the proper ID directly on the hour. If in the control room riding a football or basketball game. it's easy to insert the ID between plays or during a time out.

There's a definite, specific, legal way to give station identification: " . . . the station's call letters immediately followed by the community or communities specified in its license as the station's location . . . "

In other words, say, "This is WXXX, Smithville."

Do not say, "This is WXXX," leaving out the city of license.

Do not say, "This is WXXX, broadcasting rock and roll from Smithville." That inserts something between the call letters and the city of license.

Some FM stations use "FM" as part of the call letters, such as "WYYY-FM, Smithville." But if WXXX is an FM station, do not say, "This is WXXX-FM, Smithville." That's inserting something, "FM," between the legal call letters and the city. Picky, picky, picky. But the FCC sometimes fines for failure to broadcast full, legal station identification at the proper times.

Besides, the ID is the legal name of the station. How do you like it when someone mispronounces your name?

## How WGFR Works

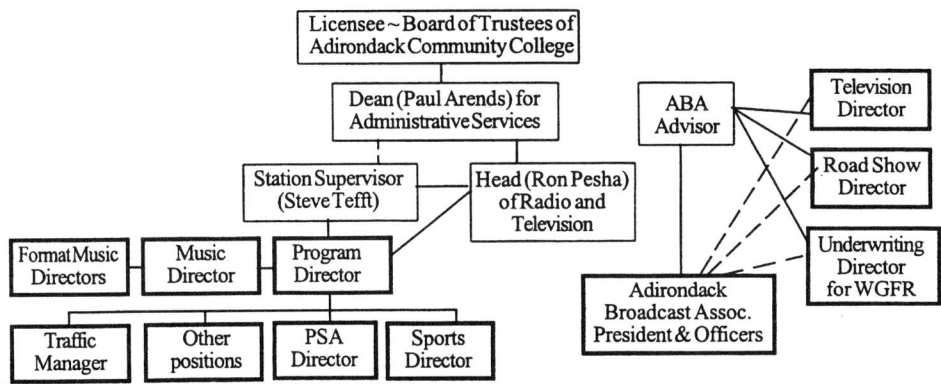

## Announcing Staff—Just About Everyone

### Who's Responsible for What

- Traffic Manager—Responsible for preparing the daily log and placing it in the control room ready to go.

- Sports Director—Responsible for sports shows and play-by-play coverage, including getting schedule of home and away games, checking out remote gear, cellular telephone, etc.

- Business Manager—Responsible for seeking underwriting for WGFR.

- PSA Director—Responsible for seeing to it that all those necessary PSAs are in the Control Room, recorded, logged.

- Music Directors—Responsible for acquiring music (buying, sending lists to CMJ magazine and recording companies), and working with PD on format and playlists.

- Program Director—Responsible for overall operation of station including seeing to it that all other positions are doing their jobs, the logs are ready, the schedule is filled, IDs are given at the legal times etc.

- Station Manager—Responsible primarily for monitoring via skimmer tapes and representing the licensee's interests in suitable program material.

- And everybody's responsible to everyone else. Radio stations . . . all businesses . . . succeed when people work as a team.

# Why Our Goal Isn't Freeform: All DJs Playing What They Like

DJs at Burlington County Community College in New Jersey have remarkable success finding commercial station jobs. Why? Because they run like a commercial station, with format and playlists. Stations know those DJs make good employees. WGFR must operate the same way.

In the conservative 90s people are concerned about wasted tax dollars. Is a college wasting tax money by allowing sandbox play rather than teaching how to get good-paying, secure jobs?

## Responsibility for Equipment

Students bear legal responsibility for the hardware you use at ACC, both on campus and that taken out for radio shows, Road Shows, television classes, etc. In using or taking out equipment and supplies you agree to return everything promptly and in an undamaged state.

## WGFR Positions

The outgoing PD with the faculty station manager solicits volunteers for all positions, interviews, and appoints a student to each job at the end of the academic year . . . about April. Most new PDs and Music Directors work over the summer preparing for the following Fall semester. Major positions are held by full-time students with a 2.00or better average. WGFR positions are important and valuable, but this is a college and academics come first.

## WGFR's Goal . . .

(the WGFR "Mission Statement") . . . is to run a real radio station, not a sandbox, competitively with the local commercial stations. We don't need a lot of rules. Rules create division. *Goals* build *consensus.* Responsible people do their jobs in spite of rules.

There are always *some* rules. We're a federally-licensed broadcast station, under FCC Rules and licensee policy. SeePage 2.

## Here's How We Achieve Our Goal

Maintaining continuity of for…,using a play list for selected so…, and expecting everyone to play responsibly… format.

- Keeping our station home reasonably neat end clean, and not defacing equipment. (You should see some other college stations!)

- Daily checking of logs to correct all entries.

- Making sure that EAS Tests are received and properly logged. Helping the community by broadcasting PSAs.

- Seeking remote broadcast opportunities using the cellular phone.

- Doing play-by-play sports. High school football games in the Fall and basketball in the winter, both from the gym and away games.

- Responsibility working together as a team, among WGFR, the Road Show, and the Adirondack Broadcast Association.

- Achieving respect from all the local commercial stations by broadcasting in a responsible, adult manner.

- Avoiding juvenile air names which would not go on commercial stations.

## Being a Pro

- Announcers refrain from negative comments about the station, staff, or anything else. Listeners want entertainment, not griping

- Disc jockeys use a mic boom and don't hold the microphone. They are announcers, not rock vocalists.

- Any business's employees answer the telephone courteously … even if a caller is rude.

- Announcers do not talk "off mic"to a person in the rear of the control room … small town radio.

- Announcers avoid making any comments about spots (including PSAs).

- "Inside" talk which only station personnel and/or friends understand is rude to your audience.

- Remember that you are on the air to entertain your public, not your buddies.

- Disc jockies play music on the air, not sound effects recordings or other production-only material.

- Clean CDs radially—that is, wiping from center to edge, never in circles!

- Announcers leave the control room neat and shipshape:

- Recordings properly filed.

- CD trays closed.

- Pots turned down and switches off.

### Picking up Faxed News

Unfortunately the news can't arrive directly at the station, and someone must pick it up each morning. Remember that a day's news not picked up wastes taxpayers' money. Do you pay state income tax? Do you want your tax dollars wasted?

WGFR was the first college station in the nation to contract with USA Today News Service.

*Attention all WGFR positions, Road Show Manager, and A.B.A. President: check your box on the WGFR desk regularly! If you're not in the habit of checking mail routinely, develop adult responsibility and get in the habit!*

### What's Appropriate

Often it's not what's right or wrong, or what is legal. It's what is *appropriate.*

What's appropriate for an urban station differs from a rural community in a conservative area . . . such as the Glens Falls market. What's appropriate on a large private university's station differs from what's appropriate at a public college.

There's nothing wrong with sexually explicit song lyrics or the so-called "four-letter words." But there are places where such material is not appropriate. One such place is a small tax-supported community college in a conservative area.

Remember that a *single complaint* to the licensee, the Board of Trustees, might make them decide that ACC doesn't need a radio station.

Remember that we compete with the local stations. *If it's unacceptable for a commercial station in this conservative market, it's unacceptable on WGFR.*

## EAS—The Emergency Alert System

### Check EAS before Sign On

1. Press SPKR once for monitored FM station, again for weather station. At least one of these stations must be heard or it is not legal to sign WGFR on.
2. Press RESET to turn off speaker.

### To Hear Weather

1. Press SPKR once for monitored FM station, again for weather station.
2. Press RESET to turn off speaker.

### To Send Weekly Test*

1. Turn on microphone and say "This is a weekly test of the Emergency Alert System."
2. Press WEEKLY TEST.
3. After ON AIR RELAY red light goes out (takes about 13 seconds), say "This was a test of the Emergency Alert System."
4. Return to normal programming.
5. Tear off printout and staple to log.

### To Send Monthly Test**

1. Press PASSWORD, then enter 9 1 1
2. Press REQUIRED MONTHLY TEST.
3. Press flashing READY button.
4. Press other flashing READY button.
5. To initiate the test, press HEADER (HDR).
6. After ON AIR RELAY red light goes out play recorded spoken message.
7. Press EOM (End of Message).
8. After ON AIR RELAY red light goes out resume normal programming
9. Tear off printout and staple to log.

### To Practice Weekly Test

1. Press PRACTICE.
2. Press WEEKLY TEST.
3. Tear off printout and discard.
4. Press PRACTICE again to turn off.

### To Practice Monthly Test

1. Press Practice.
2. Follow instructions 1 through 8 above under To Send Monthly Test.
3. Tear off printout and discard.
   *Weekly tests are run between *:30 AM and local sunset, except during a week with a monthly test. ** Monthly tests in odd numbered months run between 8:30 AM and local sunset; in even numbered months between local sunset and 8:30 AM.

## When an EAS Message Is Received

1. Drop everything and listen, even if it means dead air on your station.
2. If a test is received, tear off the printout and staple it to the station log. No other action is necessary. Do not say on the air that a test has been received.
3. An actual national emergency will automatically interrupt your station's programming and be retransmitted. You will hear it, and the ON AIR RELAY red light indicates that this message rather than your station is being transmitted. Follow the posted procedure with the EAS Checklist (see below) which may include leaving the air.

## EAS Checklist

This FCC document must be available at the transmitter control point at all times. As the operator, you must know its location. It contains information about proper procedures for EAS Alerts and EAS Tests.

## Other Emergencies

1. Your EAS unit may have been locally programmed to retransmit messages other than national emergencies. Examples include tornado, flood, hurricane, and blizzard warnings or watches and warnings, civil emergencies, and similar situations.
2. You will hear all of these messages. If they are being retransmitted, the ON AIR RELAY red light will be on, indicating that your station's programming has been interrupted.
3. Any emergency message you hear with the ON AIR RELAY red light off means that your station's programming i s continuing on the air normally.

    *As an announcer, you may view the EAS as a joke, but the FCC doesn't! The morning DJ at KSHE, St. Louis, broadcast a false EBS Alert on January 29, 1991, during the Gulf War. His employer was fined $25,000. How long do you suppose his job lasted?*

# The Tascam Digital Audio Tape Machine

DAT (Digital Audio Tape) machines work almost the same way as CD players with a few differences. Operate the DAT with the remote control and not with the buttons on the face of the machine. The remote has many more capabilities.

First, place the cassette in the machine face up with the door of the cassette facing the interior of the machine (see diagram). Always keep the drawer closed except when loading or unloading a cassette.

*DAT records time code continuously from the beginning unless you stop the tape and forward it, creating a blank section. Blank sections make searches difficult.*

After the cassette has been properly placed in the machine press the "open/ close" button. Refer to your list and find the track number of the song you want to play. Using the numeric keys, enter the track number of your song (if you make a mistake press "clear" and choose again). Now that your choice is in press the "pgm" button (the word program should light up on the display). To play your song press "start". DO NOT HIT "PLAY," HIT "START." The word search will light up and the machine will find your track.

Once the track is found it will play automatically. The track will stop on its own but you must also hit "stop" to clear the program. Press "rew" to bring the tape back to the beginning of the cassette. Play can be stopped at any time by hitting either "stop" or "pause." If you hit "stop" you must reprogram. If you hit "pause" you may hit it again to resume play.

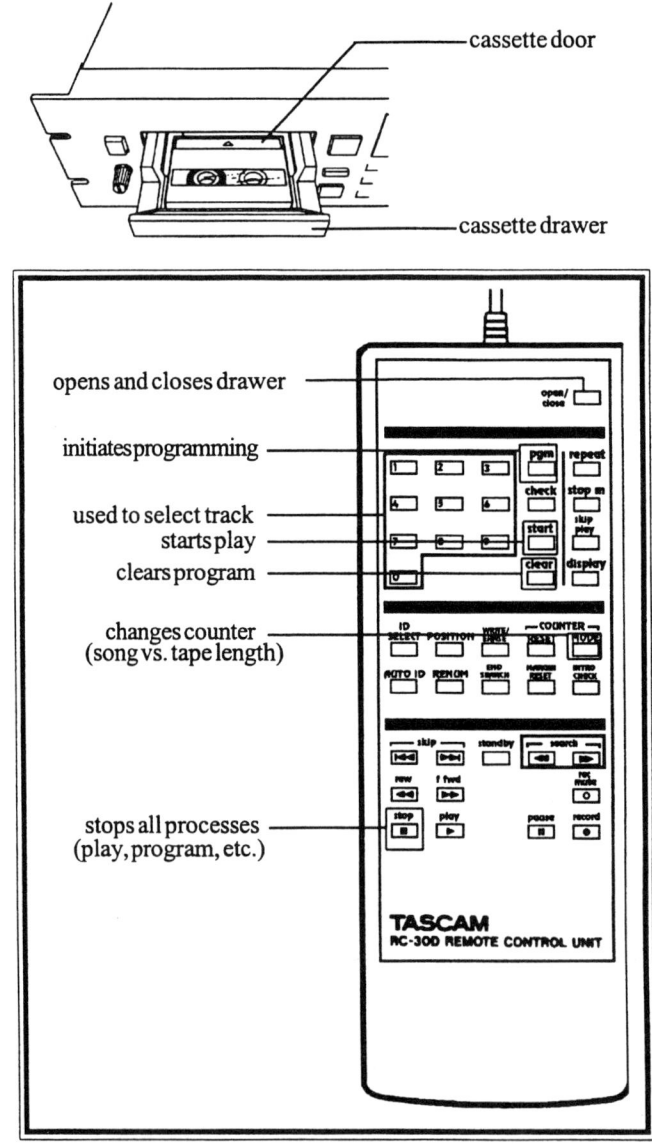

Leave a minute or two at each end of the tape blank. These are the parts of the tape most likely afflicted with dropouts.

*Avoid writing on the cassette in pencil. The manufacturer says that even the tiniest flakes of graphite can damage the machine!*

# The Sony Digital Audio Tape Machines

*To record stereo DATs from CDs, the left CD player feeds to DAT directly, not through the monaural control board. While recording to DAT you may monitor either the CD or the DAT through the control board.*

## To Record

1. Press OPEN/CLOSE and insert the cassette window side up. Press OPEN/CLOSE again. Always keep the drawer closed except when loading or unloading a cassette.

2. To star/recording at the end of are corded portion, rewind then press fast forward. The deck locates the end of the recorded portion and stops automatically.

3. Press REC and adjust the record level. Keep level between −12 and 0 db.

4  Cue the CD to the beginning. Press PLAY or PAUSE to start the DAT.

## To Insert a Sound-Muted Four Seconds Between Songs

• Just press the REC MUTE   button.

  If you do not create a sound-muted section at the beginning of a tape you may not be able to move or erase a start ID or Program Number that is recorded at the very beginning of the tape.

## Writing Start IDs or Program (Track) Numbers

• Do steps 1 through 3 above.

• Press START ID AUTO repeatedly until "AUTO" appears in the display.

• In general the numbering is automatic. For example, if you paused after recording the 5th track, "5 " shows under Pg. No. It will automatically move to 6 after you insert the Sound Muted section above and begin the next track. If you are recording from the end of a recorded portion, us the number buttons to specify the next program number.

- Now continue recording, following Step 4 above.

### Locating a Track for Playback

- Enter the track number with the number button, then press PLAY, or

- During playback, press the Fast Forward multiple times. For example, three times to go to the third track ahead, or

- During playback, press the Rewind multiple times. For example, three times to go back to the third track before, or

- During playback, press the Rewind once to return to the beginning of the current track.

### Erasing Start IDs

- Press ERASE when the start ID is displayed. This requires at least nine seconds.

### Automatic Renumbering

- Press START ID RENUMBER. The tape rewinds automatically, is renumbered, and rewound again. Renumbering may not function correctly if there is a blank section on the tape or if a start IDS exists within 10 seconds from the end of a tape.

*DAT records time code continuously from the beginning unless you stop the tape and forward it, creating a blank section. Blank sections make searches difficult. Always insert a Sound-Muted section.*

## Contests on WGFR

According to FCC Rule 73. 1216, a station "shall fully and accurately disclose the material terms" of any contests, and "no contest description shall be false, misleading or deceptive. . ."
Those "material terms" will include

1. How to enter or participate;

2. Eligibility restrictions (there may be a reasonable minimum age, for example),

3. Entry deadline dates,

4. The extent, nature, and accurate value of prizes,

5.  Time and means of selection of winners,

6.  Tie-breaking procedures, if applicable (for example, if the winner is the "fourth caller," no tie is possible).

Furthermore, these details should be broadcast a "reasonable number" of times, starting with the first announcement of the contest.

Be sure that all announcers know how to run the event, including all steps above from the first airing of the contest. Are all promo tapes, scripts, and other materials on hand in the control room? Do announcers know what to say to people who call in? What about the claiming of prizes? If winners must come to the station, what form of identification is required? Do announcers know what to do?

What about security? Is there any way which winning information might leek to listeners? It's illegal to slant winning opportunity to your friends. Emphasize security to all staff members involved—and divulge nothing to others to concerned.

Contests may not be combined with underwriting!

## Using the Nakamichi MR-2 Professional Cassette Deck

Recording: Set Tape Selector switches for Normal, High (Chrome), or Metal tapes. Keep Bias Tune Control at center detent position.

Equalizer switch: Set to 120 μ sec or Normal tapes, set to 70 μ sec for Chrome and Metal tapes.

Press Rec button to set the record-standby mode. Then press Play button to actually start recording.

Adjust Input Level controls so the +5 indicators light occasionally when using Metal tapes, or so the +3 indicators light occasionally when using Chrome and Normal tapes. In other words, Metal tapes can be recorded at 2 decibels higher level than other tapes without excessive distortion.

When the Memory switch is on, the tape stops automatically during rewind only when the counter indicates "000."

*On both the Nakamichi and the Tascam cassette recorders, set the MPX Filter switch to off when recording. (It's used only when recording in Dolby off the air from an FM stereo broadcast).*

### Using the Yamaha SPX90

*   Put pot switches for sources you want to use such as microphone or CD player to left (Audition) to feed SPX90

*   Place SPX90 switch in 3 or 4

*   Place Channel 3 or 4 pot switch to right(Program) to hear output of SPX90

### To open Computer and SoundBlaster

Before you start, take care not to change settings by random pointing and clicking. Read and follow instructions. Always use the left button on the mouse unless the right button is specified. "Double click" means press twice. To shut down follow the step-by-step instructions. Don't just press the power button. Now, to get started. . .

- Turn on the computer power (and monitor power too if that switch was off).

  When PROGRAM MANAGER appears, double click on SoundBlaster 16.

- Using the mouse, double-click on Wave Studio.

- Click on red RECORD dot to get window shown below.

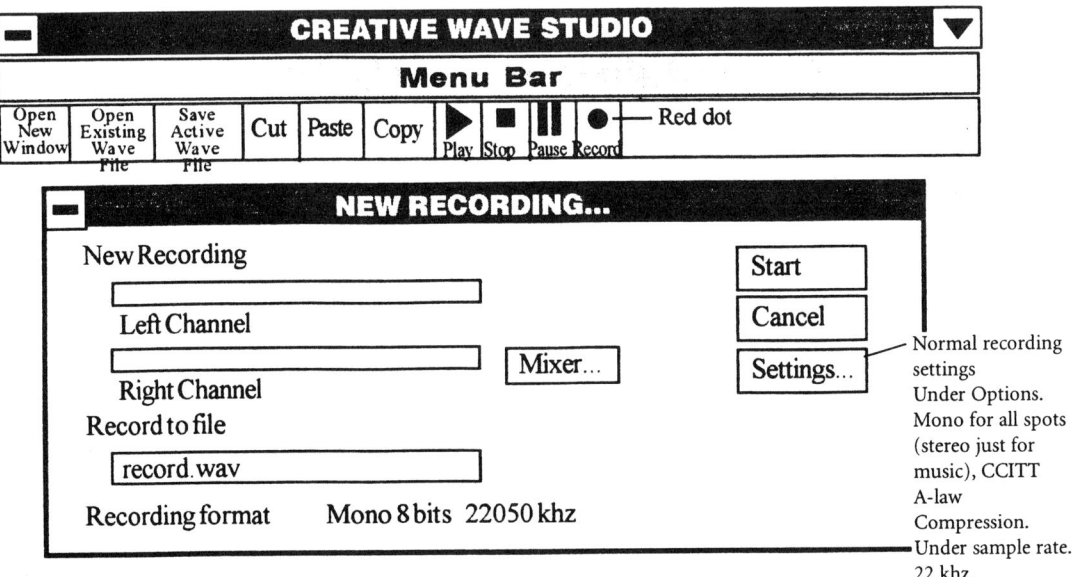

- Recording format should be Mono (for spots) or Stereo (for music) 8 bits, 22050 khz. If not, use Settings . . .

- Turn on green dot on Line In on Creative Mixer to record. Access through Mixer . . .

- Set your level, not too low (noise) or too high. If too high you clip the waveform peaks and distort the sound.

- Click on Record (the red button) and immediately start feeding audio. As soon as the audio finishes, click on Stop.

- To play back turn off green dot on Line In on Creative mixer. Play. If satisfactory, click on File . . . and Save as . . .

- If sending to the WGFR computer, transfer the r: drive. After shutdown and restart, it will appear in WGFR.

- Name the file by changing *.WAV to (for example) JOEPRO-MO.WAV. Not longer than seven characters.

## Shutting Down

To close any file, Click on the Control Menu Box symbol (the "minus") in the extreme upper left corner of that file. Then click on CLOSE. Keep going until you get to "Do you want to end your Windows session?" and click YES.

Only after reaching the DOS Prompt (C:\ or B:\ or D:\) do you turn off the computer power switch. The monitor automatically turns off with the computer. To delete the spot on your disk, Use DOS. Close Program Manager and Windows. At C: type A: and press Enter. Then type DEL *.WAV (space after DEL) and press Enter. Then turn off computer, or re-enter Windows by typing WIN and pressing Enter.

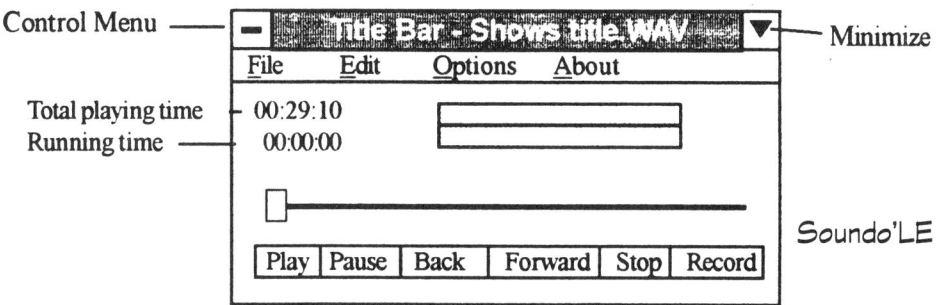

NORMAL SETTINGS
In WaveStudio click on Options
then click on Mixer Settings . . .
- Custom View(on)
  Title Bar (on)
  LED Display (on)
Preferences . . .

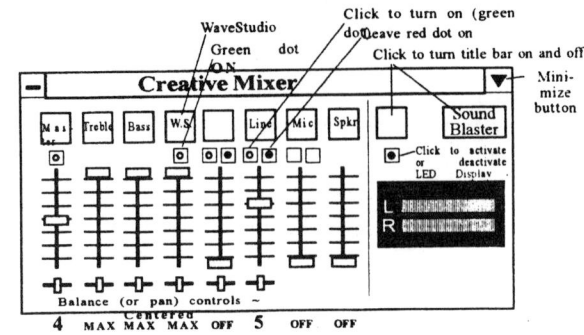

| | |
|---|---|
| X Treble | CD |
| X Bass | X Line In |
| X Wave | Microphone |
| MIDI | PC Speaker |

X Show Balance Control
X Always on Top
X Save Settings on Exit
Recording Settings . . .
Left In:
CDL off

| | |
|---|---|
| Line In | L on |
| Microphone | Off |

Right In:
CDR off

| | |
|---|---|
| Microphone | Off |

Gain: L x4 R x4 Studio C
L x! R x1 WGFR & B
AGC off
Output Gain . . . L x1 R x1

Note level controls above have seven calibration marks. Keep Master Gain at 4; Line In at 5. You *may* change bass and treble, but always restore to maximum.

**Important**
Turn on Line In (green dot on) to record.
Turn off Line In (green dot off) to play back.

Normal recording settings Under Options, Mono for all spots (stereo just for music), CCITT A-law Compression. Under sample rate. 22 khz.

---

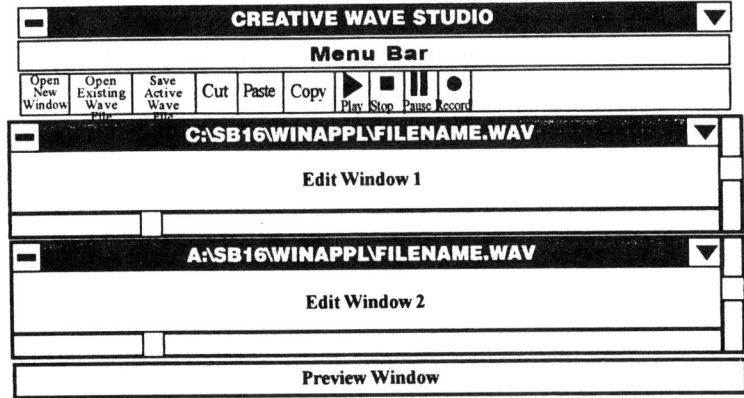

- Open Wave Studio. Click on File, Open . . . and select the desired file. Repeat for up to four files.
- Under Window click on Tile Horizontally. (Later you may prefer Tile Vertically).
- You can play any file, determined by clicking on the Title Bar of the desired Edit Window. This file also shows in the Preview Window along the bottom, with the length in seconds and the recording characteristics showing in the Status Bar just below the Preview Window.
- If the level of your file is too high or too low, alter them using Special and Amplify Volume.
- If the Recording Settings of your questions differ, use Special and Convert Format.
- Can't see all the waveform? Use the vertical slider on the right side of the Edit Window.

# The OPEN and SAVE AS Window

**File name:**

```
*.wav                          ↓
```

**File name:**

```
File name:              ↓




                        ▓
```

**List files of type:**

```
WAV files:              ↓
```

**Directories:**

c:\

```
☐ c:\                   ↓
  ☐ students
    ☐ (your name)



```

**Drives:**

```
c:\                     ↓
```

Replace the asterisk with your file name . . . 8 characters maximum, no spaces.

Be sure to show WAVfiles here. If not, scroll down to find WAV.

If Directories above does not show c:\ then scroll hereto find c:\ (use R drive only for WGFR).

Click on c:\ to find students. Click on students to find your name. If your name does not exist and you do not know how to create a subdirectory with your name, just save your file under students for now. (Find six instead for sound effects).

## Questions and Answers

**May I play a recording I've brought in to the station myself?**

Yes, but . . .

- It must be licensed to ASCAP, BMI, or SESAC.
- It must fit the format and the playlist of the time of day you intend to play it.
- It cannot be offensive or obscene in nature.
- Licensee policy does not allow the broadcast of comedy recordings.

**May I use a CD of themes from TV commercials on the air?**

"No. Absolutely not," says the National Association of College Broadcasters' legal counsel. "You cannot use any portion of a commercial (or TV show, music video, or movie) without first getting permission from either the producer or distributor of the work."

**I recorded some jingles and stingers from another radio station to use in my show on WGFR.**

No way! This is copyrighted material, and may be used only by the purchaser or lessee.

**What should I do if something isn't working in the practice studios B or C?**

Alert the faculty station manager or chief engineer. If neither is available, an advanced broadcast student may point out that a switch is in the wrong position, etc.

**Do I have to play the format?**

Yes, just like commercial stations

**How do I get a specialty show?**

Write a proposal and produce, demo tape for the show.

**How many times a week may I be on the air?**

As many times as you can handle. If you're on more than once a week and someone needs air time, you may be asked to give up an extra shift.

**What if someone is in the practice studio during the time I have it reserved?**

That person leaves on your arrival. Feel free to use a studio that is not in use, even if it is signed up during that time period. However, if the person who signed up shows up, you must leave immediately.

**What if I need help during my studio time?**

Ask a broadcast instructor. Office hours are posted for both on their office window and both are there many hours which are not posted. There are also a number of knowledgeable students who are around the station many hours of the day.

**Why can't I put telephone calls on the air?**

The college maintains a liability insurance policy for WGFR. The insurance carrier specifically prohibits the live broadcast of telephone calls. We have no choice.

**What if I make a mistake on the air, such as introducing the wrong recording?**

Just keep on going. Don 't apologize, which draws attention to the error or problem. Most listeners probably didn't notice it. Always keep on going, no matter what. As in theatre, the show must go on.

And always have something else ready to play. Avoid dead air.

**What if my relief doesn't show upon time?**

Sign off, using the posted instructions. (Legally, sign off requires station identification and a statement that the station is signing off, followed by turning off the transmitter). Signing off early is highly undesirable, but the fault is that of the announcer who didn't show up, not your fault. You may not miss a class because an announcer did not arrive.

**What if I can't come in for my WGFR shift?**

You have the responsibility for finding a replacement. Just calling in at the last moment saying you won't be in is not satisfactory. Do you think any employer would tolerate that?

**Why does the station tell me what music I can't play? This violates my freedom of speech.**

Freedom of speech belongs to the station's owner. If a member of a station staff, paid or volunteer, could dictate the content of programming to the licensee, the owner, it would violate one of the most basic American principles of this free, capitalistic society. You may play your choice of music and say what you want to say on a station which you own.

**So why can't I have some fun and send a disaster warning?**

The Federal Communications Commission's "War of the Worlds" policy bars fake newscasts or "scare" announcements. In 1991, KROQ in Los Angeles fabricated a murder confession, and KSHE in St. Louis broadcast a phony nuclear attack warning, both resulting in severe fines.

**What if an F.C.C. Inspector comes while I'm on duty?**

- Ask for identification.
- Be sure that you know where the EAS Checklist is located, how to receive and log an Alert or Test, that you know where the transmitter control is located and how to use it, that you know how to give a legal station ID, and that you have properly signed on and entered times in the log.
- Try to contact one of the broadcast instructors.
- Try to contact Dean Paul Arends, who is the liaison between the station and the licensee. Dean Arends' office is in the Administration Building, Room 119.
- Explain to the Inspector that Operator Permits end the copy of F.C.C. Rules and Regulations are filed in the faculty station manager's office
- Explain that the WGFR Public File is in the office of Dean Arends, Room A-119.

**Since WGFR is a non-commercial stations, can I list area bands and say where they are playing?**

According to Alan Myers of the FCC, "it's acceptable. However if it is overdone and a station spends a lot o f time talking up events it could be construed as affecting the non-commercial nature of the license." December 11, 1997

**What about earphones?**

In general, you provide your own earphones. The control boards in Studios B end C provide only 1/4" jacks, so to use earphones with 1/8" plugs (the "Walkman" type), you will need your own adapter. As with other personal property, A.C.C. is not responsible for loss.

**Where can I get sound effects?**

The library has sound effects CDs, checked out only to broadcasters. Your Basic Broadcast Techniques instructor provides a list of broadcast students to the library main desk at the beginning of each semester.

**Where do we get new CDs for the radio station?**

The Music Director buys them, trying for those which will remain useful for a reasonable period of time. The most valuable CDs are dubbed to DAT (Digital Audio Tape) and stored. (Students may, with permission, bring in their own CDs for dubbing to DAT). New CDs may be purchased from retail stores, but the least expensive source is a wholesaler . . . we often buy from Northeast One Stop in Albany.

**Why do I have to graduate? I don't want to take all those other courses.**

- On the average a community college graduate earns far more than a high school graduate. A university graduate does even better—even in this difficult economic era.

- The specific skills earned in college (or trade school) soon become Obsolete. The ability to analyze and synthesize—cognitive skills or thinking—which you learn in the lecture and liberal arts courses help you all your life. Skills become obsolete; basic theory never changes.

- The person who successfully completes the 21 or 22 courses necessary for graduation develops persistence, necessary for success in a career. Employers seek people with proven persistence. Getting a job often depends more on the fact that you got through college rather than on the specific courses taken.

**Suppose I get criticism, or hear criticism about WGFR?**

People of all ages, and especially young people just exiting adolescence, tend to mock and scorn that which they do not understand, lack the skill to do, or fear to try. It's okay to be offended or hurt, but don't be surprised if some people deride or jeer at your communications efforts, on or off the air. Some former A.C.C. radio students ridicule WGFR, even though they started here. Once working in the commercial world, WGFR in retrospect looks like kindergarten. But everyone starts somewhere!

**What if I have a problem and don't want to see an instructor?**

See the Division Chair, Ms. Diane Dalto, Dearlove Room D-3 17.


## Taking Out Equipment

As broadcasting is a community function, you may need to use ACC equipment including

- Cellular phone

- Remote radio mixer

- Road Show

- CDs; audio/video tapes

- Camcorder

- Remote TV equipment

- Stereo recording equipment

*Students bear legal responsibility for undamaged return of all equipment and supplies.*

# Welcome to ACC Radio-Television

On behalf of the staff and current broadcasting students at Adirondack Community College, we welcome you to the world of broadcasting.

As a student of Adirondack Community College Broadcasting, you are entering a very exciting time in your life, We, the WGFR Staff, have provided you with this handbook to help you along on a day-today basis.

WGFR was the home for many successful radio personalities such as Bob Welch, formerly with WPYX, WYLR, and other regional stations, and Tom Jacobsen, Program Director at WYLR in Glens Falls. Recent students who began at WWSC/WYLR while still students include John Culver, Mike Dubray, Josh Greene, and Jason Wentworth. Brian Delaney and Joe Donahure among others are at WENY/ WSTL, Jeff Davis at WKBE, and others at WBZA/WAYI. One Adirondack broadcast student went into acting. Scott Valentine, who played Nick on the hit TV show "Family Ties," was a broadcasting student and worked in the WGFR studio. And two members of WNYT, TV Channel 13 in Albany, started here: Steve Scoville and Mark Mulholland.

This handbook outlines the programs and operating procedures of WGFR and identifies the staff and each of their responsibilities, WGFR policies and procedures, rules and regulations, and student participation, which together serve to make a vital social and academic growth for everyone.

It is our hope that this publication will be used as a continuing reference source. It is our belief that through the understanding of its contents, it will offer each student his/her greatest opportunity for uninterrupted academic and social growth while here at Adirondack Community College.

## The WGFR Radio Studio

WGFR is located in the rear of the Academic Computing Center, on the lower floor of the Library Building. Located next to the WGFR studio are practice studios B and C. Both B and C are for lab work and their use will be discussed in the section of student responsibilities.

The office of the broadcasting instructors at A.C.C. is located between studio B and studio C and their office hours are posted on the window of the office.

We, the staff and current students in broadcasting, welcome you and offer our help and experience to help make your time with us exciting, memorable, and educational.

## WGFR and ACC Broadcasting History

The radio-television broadcasting program at Adirondack Community College continues its consistent effort to grow and expand, to spark and enhance creativity, to open up the world of mass communications for hundreds of our country's future broadcasters.

To list all of the programs' accomplishments in narrative form would require considerable space. We summarize events since WGFR went on the air.

January 17, 1977. A.C.C. radio station WGFR goes on the air under the manager Ron Pesha and the students of the Mountaineer Broadcasting Association, instituting a variety of local news, interviews, public affairs shows, community events, educational programs, extensive sports coverage, along with a wide range of music from classical to jazz, and from country western to rock, with a transmitter located on campus. The frequency is 91.9 megahertz.

- 1979–1989 WGFR increases efforts toward public interest in producing extensive localized programming with educational courses from both A.C.C. and Southern Illinois University.

- Station publicity expanded through T-shirts and bumper sticker promotions.

- Up-to-date programming listings entered weekly in the Post Star newspaper, along with live readings twice daily of "Upcoming A.C.C. Events" open to the public.

- First ever live remote broadcasts at WGFR with home games of Queensbury High School football.

- Broadcasting rights received by WGFR for New York State High School Basketball playoffs at the new Glens Falls Civic Center.

- WGFR broadcasts 44 Double A Professional Baseball Glens Falls White Sox home games from East Field all summer of 1985, and ten games from Burlington. The play-by-play personality for most games is Brian Delaney, now with WENU.

- UPI contract obtained by WGFR offering up to the minute news with 5 minute newscasts on the hour with a 5 PM 20 minute news round-up and an additional 10 minutes of sports news.

- 1983 WGFR moves its 10 watt transmitter from A.C.C. campus to the roof of the Continental Insurance Building in downtown Glens Falls. Location donated by Continental Insurance with the new transmitter paid for by a $1000 grant from the Glens Falls Foundation and by $1700 from the Faculty-Student Association.

- Frequency moved to 92.1 in the commercial FM band.

- 1984 WGFR broadcasts through solar power, participating in a project carried by the Atmospheric Sciences Research Center of the State University of Albany. The main control board for

WGFR was designed to use photovoltaic system by broadcasting instructor and station manager Mr. Ron Pesha.

- 1985 WGFR . . . 92 Rock broadcasts high school football games each Friday night from East Field' Glens Falls. WGFR broadcasts sole coverage of Northern Adirondack League Super Bowl games.

- June 8, 1989 WGFR receives national attention when USA Today calls the station ( not any area commercial stations) for a comment on The Who rehearsing at the Glens Falls Civic Center.

- 1989 Students use WGFR to raise over $ 1300 to benefit the victims of the California Bay Area earthquake, sending the money to a college station in the area for forwarding to the Red Cross. An article about this project appears in College Broadcaster magazine.

- 1990 Broadcast students win the coveted Presidents Cup in conjunction with A.B.A., the Adirondack Broadcast Association.

- 1991 Broadcast students again with the President's Cup, shared with New Horizons club.

- 1992 Broadcast students start producing a syndicated radio show for the Adirondack Mountain Club heard on noncommercial and commercial radio stations around New York state.

- 1993 Broadcast students win the President's Cup for the third time.

- 1994 TV broadcasters not only do production for the 7th Annual Prospect Child & Family Center Telethon, but put on a live cable TV debate between candidates for Queensbury Town Supervisor (sponsored by the League of Women Voters), and also produce a five-hour Telethon in April to benefit burned out ACC student Janet Truelove, an all-student in-college show. Cellular telephone remotes commence with live coverage of the Robert Hutchinson Child Care Center dedication featuring guest speaker Matilda Cuomo, wife of the Governor. Broadcasters win the President's Cup for the fourth time, three times an exclusive win, a record at ACC.

- 1994–95 Change to 94.7mhz forced by religious station applying for 91.7. Justin Chabot is WGFR Program Director. Broadcasters compete in the community with commercial stations, hosting the Glens Falls Downtown Christmas Festival. WGFR broadcasts live from the November Aids Awareness Night. For the fourth year, WGFR helps with the Regis Hairstylists' Breast Cancer Awareness Drive. Broadcasters present a panel to a national audience in a packed room at the National Association of Broadcasters Conference in Providence. The ABA, Megan Rocque as President, enters a float in the South Glens Falls Christmas Parade, playing seasonal music and putting decorations on a tree from a flatbed truck bearing the sign, "Decorate Your Tree with WGFR."

During the Spring semester, the SuperJam Rock Concert fills the theatre and sells $2607 in tickets and concessions, all donated to Prospect Child and Family Center. The 8th Annual Prospect Telethon looks better than ever. And the Adirondack Broadcast Association wins the President's Cup for the fifth time, honored to share it with Circle K (with which the Road Show has performed dances).

- 1995–96 Steve LaForty was PD for the first semester, Adam DeVoe for the second. Craig Porter was ABA President. No, we didn't win the President's Cup this year, but accomplished a lot.

- Two Rock Concerts. Holiday Ball grossed about $540, SuperJam $1467.

- Toys for Tots collection and music in Aviation Mall totaling $290.

- Nineteen Road Shows including open-air downtown Glens Falls Tree Lighting Festival and three shows at Lake George Bowl.

- Basketball with Darla Belevich's Lady Timberwolves, March 20. Thanks to them $ 172.50 was donated to Prospect Child and Family Center. ABA won!

- Twenty-six ABA students attended the annual conference of the National Association of College Broadcasters (NACB), the world's largest student media organization. WGFR staffers presented a panel on formatting a station. About 275 schools attended this national conference averaging only about four students per school. Our 26 was the second largest delegation of any school in the nation, exceeded only by SUNY Oswego.

- Videotaped Tough Man at the Civic Center, under Sports Director Brendan Greenwood, selling the tapes.

- Took a trip to ESPN headquarters, Bristol, CT, on May 3.

- 1996-97 Adam DeVoe was PD for the first semester, Brendan Greenwood the Sports Director, and Darlene Erck ABA President. J. R. Hanna was Road Show Director, Micha Conover the Asst. Director.

- WGFR broadcast three home football games live from South Glens Falls High, and for the first time ever videotaped a game, the Oct. 5 clash with Glens Falls.

- WGFR celebrated its exact 20th anniversary on January 17,1997 with a banquet attended by some fifty people.

- Charles Gelarden was PD in Spring '97.

- Fall 1998 Darlene Erck was Program Director, Ryan Tumbaugh M.D.

## College Radio Stations

*About 80 community y and junior colleges in the U.S. have student radio stations, 8 in New York*

*About 160 high schools, public and private, have student radio stations, 8 in New York.*

## Getting Jobs

*All local radio stations have employed ACC students and graduates, most at WWSC/WYLR. Stations actively seek employees. For example, WCKM urges qualified ACC students, including females and minority individuals, to apply.*

# Underwriting on WGFR

WGFR is non-commercial, and cannot sell advertising or even mention commercial businesses and products except under narrow legal limits defined by the FCC. However, we can solicit donations to help the station(promotions, gifts of recordings, equipment, or money, contest give away merchandise, etc.).

**As a non-commercial educational station, can WGFR use company slogans in on-the-air underwriting announcements?**

From College Broadcaster magazine, April/May 1992: "If the slogan does not contain any comparative or qualitative language, and cannot be construed as promotional, the identifying slogan can be used. The rule here is that you must make a good faith determination as whether or not the slogan sounds too much like a commercial, and less like an underwriting spot. For example, the slogan, 'GE: We Bring Good Things to Life,' is permissible because the slogan is not product-specific, and the language is not persuasively promotional. However, the slogan, 'Metropolitan Life Insurance: Get Met, It Pays,' is not permissible because the slogan is definitely promotional."

**May I use a music bed and production for an underwriting announcement?**

As pointed out by the National Association of College Broadcasters' legal counsel, the F.C.C. takes the position that underwriting announcements must be "bland." Any information broadcast on WGFR which refers to a commercial business directly or indirectly must be just that—informational. For example, a listing of current venues for area bands' performances maybe broadcast, but only with royalty-free "production" music as a bed. The music may be pleasantly upbeat, but not hard-driving beds as typical in "hard-sell" commercials. All such announcements must be informational, not promotional. (From a telephone conversation with Kristine Hendrikson of the National Association of College Broadcasters on November 21, 1996. Hendrikson had just talked with NACB legal counsel Cary Tepper).

**What must I avoid in preparing an underwriting announcement?**

1. Use no value descriptions (such as adjectives), or qualitative or comparative words. Just pizza, not "good," or "great," or "mouthwatering," or "tasty" pizza or "the best."

2. Use no call to action terms, as is usually done in commercials. Don't say "go to," "come on down," "get your," or "call this number."

3. Do not mention sales, discounts or contests in any way.

**How long can underwriting announcements run?**

According to National Association of College Broadcasters legal counsel, "If an underwriting announcement takes longer than 15 seconds to announce, it is likely (according to the FCC) to be too promotional" (College Broadcaster, Vol. 8 No. 1, p. 17) Also, we may not guarantee any specific amount of air time in exchange for donations.

**What's a *safe* way of preparing underwriting messages?**

"This program was made possible by—" or "Music on this program was made possible (in part) by—. " Avoid use of the word "sponsor. " Use only royalty-free production library music for a bed, never regular music recordings because this violates copyright law.

Such announcements maybe made at the beginning and/or end of a program, or at any natural break in the program (such as between recordings). You may not interrupt a program for any underwriting announcement. Remember that WGFR policy requires a written agreement with each client for all underwriting, which must be signed by the station manager. Students lack legal authority to sign agreements.

**Since WGFR is a non-commercial stations, can I list area bands and say where they are playing?**

According to Alan Myers of the FCC, "it's acceptable. However if it is overdone and a station spends a lot of time talking up events it could be construed as affecting the non-commercial nature of the license." December 11, 1997

**Suppose WGFR sponsors a night at a local club, promotes the band, and receives the cover charge while the club takes drink sales and all other profits?**

"Unless you receive 100% of the profit, you cannot promote it on the air." Alan Myers, FCC, December 10, 1997

**The Adirondack Broadcast Association and Underwriting** recognizing that WGFR is federally licensed to the Board of Trustees of ACC, end that federal law requires all underwriting earnings to go to the Station, student participants in ABA underwriting activities on behalf of WGFR are not subject to club vote and/or appointment by Club officers but are under the direct supervision of the Club Advisor or another employee of ACC who has been so designated by the licensee through the Dean of Administrative Services. Merchandise and services received through underwriting will be delivered directly to WGFR in a timely manner.

Underwriting monies will be deposited in a separate FSA account, designated 6055A, and the Club Advisor or other ACC employee in charge will see to it that such funds are transferred in a timely manner to WGFR for appropriate station use. It is understood that the Division retains the right within federal licensee guidelines to veto any purchase of hardware which it deems inappropriate forth instructional direction of radio-television. Adopted by Student Senate December 3, 1997.

## Free Cash Scholarships

### The ABA Scholarship

Traditionally the Adirondack Broadcast Association earns enough money to fund two annual $400 scholarships, applicable toward tuition. The requirements:

- CQPA of 2.5 or better

- Plan to attend ACC next year.

- Radio-TV major.

    If no Radio-Television major applies, the scholarship(s) may go to Communications and Media Arts majors, and finally to any major.

### The Harron Communications Scholarship

Harron also funds one annual $400 scholarship just for Radio-Television majors. The requirements above apply, plus the applicant's address

- must be in a Harron cable TV service area.

*Scholarship Application Week comes in March*

## WGFR 92.7 FM

640 Bay Road Queensbury, NY 12804
743-2311 Manager: 743-2300 extension 567

As a non-commercial radio station with limited funding, WGFR gratefully acknowledges donations for acquiring equipment and for under writing the costs of operation. The Adirondack Broadcast Association takes pleasure in acknowledging such donations and underwriting by means of on-the-air announcements on Station WGFR, within the limitations imposed by Federal Communications Commission Rules and Regulations. We may identify donors by business name, location, and product lines, but we may not use qualitative language such as adjectives. All language must be "value-neutral." We may not urge people to go to the place of business. We may not mention contests, sales, or prices in anyway, even indirectly.

For example, we may broadcast a recorded announcement which reads, "Some of the expenses of broadcasting have been made possible by a grant from McWendy's, on the corner of Main and Broadway in Any city, offering the Large Mac and Chicken McKings." Such an announcement must be delivered in a conversational manner without "hard sell" or hyperbole. Announcers are cautioned to play such announcements only when logged, and without comment. We are not allowed to guarantee a specific amount of airtime for these announcements, and each announcement should not exceed 15 seconds, according to legal counsel. The underwriter is cautioned not to provide special services or prices to any individual from the station for any reason. Under federal law, this is not allowed.

Your announcement will be recorded, will use the term "underwritten by" (not "brought to you by," "sponsored by," or any term other than "underwriting"), and will read this way: (copy may be attached rather than written here)

It will be logged to play on the air beginning on this date _ _ _ _ _ _ _ _ _ _ _ _ _ _ and ending on this date _ _ _ _ _ _ _ _ _ _ _ _ _ _ _ _ _ _, and/or as part of this program _ _ _ _ _ _ _ _ _ _ _ _ _ _ _ _ _ _ _ _ _ _ _ _ _ _ _ _

Liability limited to refund of the cash amount of payment.

Salespeople:

_____
WGFR station manager

_____
Client Signature

# Receipt
*WGFR's Copy*

Received of  _____ $ _____ (Cash/Check)

_____
Signature of WGFR Account Executive

# Your Instructors and Station Supervisor

*alphabetically*

## Mr. David Kieserman

*Instructor of Speech*

Kieserman joined Adirondack Community College in September, 1996, and took his first teaching job in September 1958. He was a 20 year old graduate of Montclair State Teachers College in New Jersey and as a new teacher in the Atlantic City High School, he was stopped by at least four older teachers in the hallways wanting to know where he was "supposed to be." They thought he was just an overweight senior who was cutting class.

Since then he has taught at the University of Illinois (Urbana campus), the University of Chicago Laboratory School, George Washington University (D.C.), Skidmore College, Nicolet Technical College (Wisconsin) and Stern College for Women(New York City). During these years and beyond he has maintained an acting career which has included films, television (series, soaps, and commercials), and radio and stage productions. He has traveled to India (on a Fulbright Grant), Japan (to appear in Little Shop of Horrors), and Greece (to study ancient theatre)just to name a few ports of call. While living in New York City (1980–1992)) he worked for the New York State Council on the Arts and appeared off-Broadway in The Fantasticks, the longest running musical in the world. His film credits include Broadway Danny Rose, The Way We Were, Purple Rose of Cairo, Ghost Busters, The Film of Daniel, Single White Female, Green Card, Hair, The Flamingo Kid, and several others. TV credits include Law and Order, Nurse, The Rage of Angels, Edge of Night, Spencer for Hire, My Old Man, and Guiding Light.

David has taught speech and diction wherever he has taught as well as on active duty with the United States Air Force. Following military service he earned his M.A. at the University of Illinois and finished course work for his Ph.D. at New York University. He lives in Saratoga Springs with his wife, daughter, two family dogs and four computers. Since 1985 David has operated his own tax preparation business, ACTORTAX, serving American actors around the world.

## Mr Ron Pesha

*Associate Professor of Broadcasting*

Pesha joined Adirondack Community College in 1976, placing WGFR on the air on January 17, 1977. He is advisor to the A.B.A., the Adirondack Broadcast Association. B.S. from Skidmore. M.A. (in Humanities, emphasis on literature) from California State University.

- Appointed in March, 1993, as a member of national Faculty Advisory

Board of the National Association of College Broadcasters.

- Advisor of the Year, 1985, and again in 1993.

- Teacher of the Year, 1989, and again in 1995.

- The President's Award for Excellence in Teaching, 1991.

- State University of New York Chancellor's Award for Excellence in Teaching, 1993.

   Mr. Pesha is the first, and as of 1995, the only person in the history of ACC to win all four faculty awards.

- Seventeen years in commercial radio in announcing, news, program director, and some sales.

- Two years in maximum-power TV as director, on-camera talent, and photographer.

- Assistant manager of KFOA (FM), Honolulu, 30,000 watts.

- Disc jockey and chief studio engineer with KBCA (now KGGO), Los Angeles, ranked (by Pulse) as sixth to second in the market in all day parts.

- Working chief engineer at many stations: designed and fabricated two full size stereo control boards at KFOA.

- Numerous articles in Broadcast Management/Engineering, College Broadcaster, Journal of College Radio, Feedback, and Broadcast Engineering. Regular columnist for "Roots of Radio" in Radio World.


## Ms Jessica Redmond

*Instructor of Broadcasting*

Redmond is Traffic Manager at WWSC/WYLR, Glens Falls, and is on-call for news and on-air work as well. Before joining WWSC/WYLR, she was Office Manager/News Fill at WKBE, Glens Falls. Redmond was Midday Announcer/Business Manager at WMJR/WBZA, Glens Falls for three years.

   Redmond's broadcasting career includes stops at Emmis Broadcasting's WQHT-WFAN in New York City, and New City Communication's WSYR/WYYY in Syracuse. She was Promotion Director/Weekends at one of the nation's first FM-FM simulcast radio stations: WSHQ/WSHZ, Albany.

   In addition, she was an Account Executive at WQQY/WKAJ, Saratoga Springs and worked on the air at WVICZ AM/FM, Albany.

Redmond earned her B.S. Degree from the S.I. Newhouse School of Public Communications at Syracuse University. She was Station Manager at WJPZ, a highly respected student owned and operated FM station at Syracuse.

She has been active in many nonprofit organizations, serving on the board of the Capital District Chapter of the Arthritis Foundation and as Vice President for Individual Development in the Saratoga Springs Jaycees. She has spoken about broadcasting to many groups including the Voluntary Action Center and the Girl Scouts.

## Mr. Steve Tefft

### WGFR Station Supervisor

Our Station Supervisor was also the first DJ on the air when WGFR started up on January 17,1977. Why? Because Steve was an ACC student. He started in commercial radio at WPIS, Ticonderoga in May, 1977, as Sunday morning sign-on announcer.

His first full-time position came at WWSC in November l977. Later at WLAN, Lancaster, PA, Steve held down the 7 PM-midnight slot, Arbitron rated #1 in that top-75 market. Back to the Glens Falls market, Steve worked at WENU/WSTL as station manager 1986–90 and also afternoon drive at WMJR/WBZA.

Back at WWSC as PD, some of Steve's credits include:

- Hosting the "SpeakUp" talk show, interviewing local and national personalities and newsmakers.

- Creator and host of the Thursday morning "Golden Oldies" show.

- Developing When Radio Was, Imagination Theatre, and other nostalgic radio shows.

- Helping organize and promote local fund raisers for the Red Cross, Salvation Army, Caritas, the Glens Falls Association for the Hearing Impaired, etc.

## Myths About Equipment

Costly CD players sound better than cheap ones. So advanced is compact disc technology that the quality of sound varies hardly at all from the most expensive to the least expensive. Better CD players offer extra features (programming, readouts, etc.) plus more rugged construction for longer life. But they sound the same.

Big stations are louder than GFR's 10 watts. Station power determines range, not loudness. Any large market includes many stations ranging from 10 watts up to 25,000 and 50,000 watts. So long as your receiver picks up a full-strength signal, they all sound about the same loudness. A 10,000 watt station is not ten times as loud as a 1000-watter.

New speakers are better than old ones. Speaker technology continues to improve, with good sound from smaller cabinetry. But you get what you pay for. A high quality 40-year-old speaker is preferable to an inexpensive modern speaker, though it might not look as good, or be unwieldy in size and weight. Speakers rarely wear out. If not overdriven or otherwise mistreated, a speaker may easily last fifty years, perhaps a hundred.

New microphones are better than old ones. Just like speakers, you get what you pay for. The large "ribbon" or "velocity" microphones of the 1930s reproduced sound very well . . . not as good as today's best mics (at several hundred dollars)' but equal or superior to modern inexpensive microphones. A microphone is even less likely to "wear out" than a speaker. Cared for, a 1930 microphone works fine, and probably will in the year 2030.

Digital recording sounds better than analog. Usually. Digital recording dubs (copies) more accurately and offers fewer hurdles to top-quality sound, along with a much quieter background than either analog tape or long-playing records. All vinyl records are analog. "Digital" merely means that the master tape was in digital form . . . desirable, but many digital advantages are lost when transferred to vinyl. The very best analog recordings (on cassette tape and vinyl records) sound very good, though without the velvety silent background of true digital recording on CDs or DAT.

The more you spend, the better the sound. Not always true, because you may be buying features rather than sound quality. And once top sound quality is attained, a small improvement requires much more money.

Big stations sound better than WGFR. WGFR uses professional CD and cassette players and DATs. We record spots digitally on computer hard drives, a technology which many commercial stations do not yet have. Even inexpensive modern amplifiers and control boards feature flat frequency response and distortion figures under 0.1 %. We have a state-of-the-start FM transmitter. The only weak link in the chain is the line between the control room and the transmitter which does roll off the very highest frequencies, but we send audio through fewer processing steps than most commercial stations. The only major deterioration in WGFR sound originates with operators overdriving the equipment (not watching VU meters) and causing distortion. Of course WGFR is mono, while nearly all other stations are stereo.

# So what's Special about A.C.C.—"Bay Road Tech?"

## Quite a Lot, Really—Read On

- For 1996–97 ACC was the only SUNY community college which did not increase tuition. The average increase was 6.3%. Preliminary information indicates that ACC features the lowest full-time and part-time tuition in the SUNY system and perhaps the lowest of all colleges and universities in New York State.

- In 1992, Adirondack Community College won a highly prestigious national award for retention (keeping students through graduation)—the only institution of any size in New York State, end the only community college in the entire nation to receive the award.

## Radio-Television/ Telecommunications Students at A.C.C Do Much.

- Operate an on-air federally-licensed radio broadcast station.

- Function as a corporate television production facility, creating videos actually used by local organizations and seen by the public.

- Perform all production for the annual Prospect Child and Family Center Radio/Telethon, seen live on regional cable systems (Harron and Glens Falls Cablevision), and heard live on WWSC radio.

- Attend the annual conference on the National Association of College Broadcasters at Brown University in Providence, R.I.. . . a student event attended by students from 175 colleges and universities throughout the nation. See sidebar, right. The Conference takes place in November, and while not officially sponsored by the college (you are on your own), you are encouraged to attend with Mr. Pesha assisting in calling in Conference reservations. Traditionally, ACC broadcast students typically make up the largest delegation from any community college in the nation. In fact, we're a larger group than all but one or two of the big universities! ACC students presented a panel in 1994, and is scheduled to do so again in 1995. In March, 1993, Mr. Pesha was appointed for three years to the National Faculty Advisory Board of the NACB. If you attend, you are responsible for all your expenses, travel arrangements, accommodations, etc. ACC bears no responsibility whatsoever. Providence is a large, potentially dangerous city, so stay in groups. See the "Going to Providence" page for more information.

### Former Students Talk about ACC's Telecommunications

Graduate from SUNY Oswego, now working for a trade association dealing with post-production (editing) companies: "I loved the hands on experience . . . you also need the written text book material to understand the basics . . . together will best prepare you for the job you wish to do once you've graduated . . . I learn best by doing, so the classes where I produced and directed and was a crew member were an important part of my education."

Graduate of Emerson College, Boston, now with CBS-affiliate WHDH-TV: "Emerson accepted all my credits in television, radio, speech and voice from ACC. This means quite a bit that such a prestigious college would accept junior level courses as 'equal' . . . After my first year (at WHDH) as a traffic assistant, I was promoted to National Sales Coordinator. I deal with 13 rep offices in the U.S. Needless to say, I've put my speech classes to good use!"

### ACC Transfers are #1

Overall, more ACC graduates transfer successfully to four-year universities and stay to finish bachelor's degrees than from any of the 29 other SUNY community colleges.

### The NAC National Association of College Broadcasters

A.C.C. and WGFR are charter members of this student-based organization (We also belong to the Broadcast Education Association, but this is a faculty-based organization). The annual convention of the NACB, underwritten by CBS, Fox, and Turner Broadcasting among others, brings major nationally-known radio and television professionals to present workshops for students. The publication, College Broadcaster, featured WGFR and ACC six times from 1990 to 1995 . . . quite an accomplishment for a periodical which goes to the largest universities in the nation.

## The Big Broadcaster Events

### ABA

*September*

> Order Halloween toys.
> Contact Grammar school classes for Halloween art.
> Reserve theatre (H-108) for SuperJam.

*October*

All-Club "Safe Trick or Treat" for kids in the Lounge. (Outgrowth of an ABA event.) Buy small toys; solicit school children's art work; build booth.

*November*

NACB Conference Thursday–Sunday, 2nd weekend in November Started going (as charter members) in 1988

South Glens Falls Christmas Parade. Second Sunday in November. Cell phone live on-air coverage possible. Our first float in 1994; missed '95.

*December*

Dart board sales.

*Spring*

Basketball game vs. Lady Mountaineers, profits to Prospect School.

*March*

SuperJam Rock Concert. A Friday night. Started in 1988

*April*

Dart board sales.

## Road Show

*Summer and All Year*

We need money, so summer shows are possible.

*September*

Orientation Day Play music for new freshmen.
Taste of the North Country, Glens Falls City Park. Play music in non-profit area; interview people. Sponsored by Kiwanis Club.

*October*

Halloween Party at South Street Center.

*November*

Glens Falls City Holiday
Festival and Tree Lighting. Anchor music and live performance sound. First Friday in December. Contact Joann Hicks, Queensbury Town Office. Started in 1997.

*December*

Toys for Tots Holiday Music in Aviation Mall Road Show. Started 1995. Contact Marine Corps Toys for tots (Joseph Fiore, 7922687).

Teresa Tabor, Mall office. Requires insurance statement from office of Dean Arends sent to Mall before Thanksgiving. In 1997 did it at Wal-Mart.

Fort Edward Road Show "Operation Santa" Mrs. Esperti, 747-9033

*February*

Valentine Party at South Street Center.

*May*

End of month party at South Street Center.

## WGFR

*August*

Washington Co. Fair, cell phone live broadcast from ACC booth. Requires D Jon duty at WGFR control room.

Contact high schools for rights to football broadcasts.

*Football Season*

Live coverage of high school games, cell phone. Away games possible.

*Basketball Season*

Live coverage of ACC games. Away games possible using cell phone.

*December*

World AIDS Awareness Day December 1. Cell phone remote coverage, interviews. Started in 1994; missed '95.

Alumni Week Call alumni and ask them to run WGFR for an hour

*January*

Jan. 17 WGFR anniversary (went on air in 1977).

*May*

Alumni Week Call alumni and ask them to run WGFR for an hour

## Television

*August*

Contact high schools for rights to football telecasts.

*Football Season*

Videotape high school games for Harron cable.

*Basketball Season*

Videotape ACC games.

Away games possible.

*October–May*

Adirondack Arts, monthly cable show needs volunteer crew and producers

*December*

South Glens Falls Holiday Parade videotape for cable. Call Evergreen Bank. Started 1996.

*March or April*

Telethon for Prospect Child & Family Center. A Sunday late March or early April. Started in 1987

*May*

Glens Falls Memorial Day Parade videotape for cable. Call Mayor's office.

Student events in Aviation Mall and some other private venues require permission plus insurance coverage from ACC.

## The A.B.A–Adirondack Broadcast Association

**What is the A.B.A?**
- The radio-television students? Organization—your club.
- As other student organizations, the A.B.A. functions under the F.S.A.— the Faculty-Student Association of A.C.C.
- Funding comes from your Student Activity Fee.

**Who belongs to the A.B.A?**
- Radio-television students at ACC, and other students eager to help with A.B.A. projects such as SuperJam, the annual rock concert.

**Must I join A.B.A?**
- No, but it's about the quickest way to become involved in meaningful radio-television projects. Everyone feels uncomfortable at the first meeting, but you're quickly involved and participating because you are needed.

 **How do I join?**
- Just come to the meeting each Monday at 12:45 (following the WGFR station meeting which begins at 12:30) in Room L-102.

**Who runs it?**
- Students run it, electing a President, Vice-President, Secretary, and Treasurer near the beginning of each semester(officers may be re-elected). The Faculty Advisor, Mr. Pesha, is only an advisor.

**What does the Adirondack Broadcast Association do?**

Well, last year the A.B.A. did:
- On-air remote broadcasts from Aviation Mall assisting national fundraising for cancer research.
- Handed out "trick-or-treat" toys to A.C.C. Child Care Center youngsters and to guests at Cue and Curtain's annual "Haunted House. "

- Stereo-recorded concerts by the Adirondack Community College music department.
- Purchased Christmas gifts for two underprivileged local children.
- Planned, promoted, and carried out two live rock concerts in the A.C.C. theatre, including the major annual event, SuperJam.

# SuperJam

## What Is SuperJam?

- Charity rock concerts, planned and carried out by radio-television students under the aegis of the Adirondack Broadcast Association.

- Proceeds go to Prospect Child and Family Center, the local facility for handicapped children, located on West Aviation Road in Queensbury.

## SuperJam Problems

Broadcasters mounted two hugely successful SuperJam Rock Concerts in the Springs of 1990 and 1991. The large crowds and substantial profits to charity undoubtedly contributed substantially to the winning of the President's Cup for those two years.

Then there were four failures in Springs and Falls. Crowds were small, and they did not meet expenses . . . in some cases, losses were severe. Three were poorly organized and planned. The fourth, Spring 1993, was well planned, but extremely expensive and it failed to draw even a half-full house. Then on March 24, 1995, combined door receipts and concession sales equaled $2407.66.

## Formula for Success

- Low expenses

- Big audiences of high schoolers

    WGFR appeals mostly to a young audience. These people are the only ones who come in large numbers. They will come if the know the bands. And they are too young to frequent bars and other places where live music is heard in the area.

    The bands don't have to be good, just need to be popular with kids. Two of the bands for the immensely successful '95 SuperJam were from high schools. They helped publicize the event at their schools, with posters given to them for distribution.

    **The successful rock concerts utilized bands who played free.** After all, it is for charity. We got in trouble when we tried to hire expensive bands.

We must also keep lights and sound costs low. One concert used the theatre's lights, augmented by the flashers from, the WGFR Road Show. This was free. Another concert obtained sound for a maximum of $100.

With fixed costs of almost $300 for security and janitorial service, we must keep total expenses under $500., so even a small crowd will break even.

*For more details, see, "Responsible Concert Planning. "*

## The President's Cup

This award, given annually to the school year's outstanding student organization, was won by the Adirondack Broadcast Association in 1990, in 1991 (shared in '91 with New Horizons, the adult returning students' club), and was sole winner again in 1993 and 1994. In 1995, ABA was honored to share the cup (our fifth win!) with Circle K, with which the Road Show has performed several dances. The primary criterion for winning the President's Cup appears to be earning money for charitable donations. Thus, allow over head high receipts SuperJam is a key. But new people can add fresh ideas. What's your idea?

## Road Shows

For years, broadcasters at A.C.C. have hired out as disc jockeys for junior high and high school dances, parties, wedding receptions, etc. Once we played Bolton Landing's Sagamore Hotel night club for two nights in a row. We did a wedding reception aboard The Horicon on Lake George, and another at the Queensbury Hotel. And we do parties regularly as the Warren-Washington County Mental Health Association's South Street Center—they love us there!

It's a wonderful opportunity to get real experience, performing before an actual audience.

Many shows are done gratis or for a reduced rate for worthy charitable organizations. Other shows earn money which goes into a special Broadcast account, which buys recordings for WGFR and occasional other needs and charitable donations.

## Play-By-Play Sports

Interested in sports? Broadcast live basketball from the ACC gym, mid-November to Spring. See the Page about doing Play-By-Play. And other sports using the cellular phone (see next paragraph).

## Remote Broadcasts

WGFR's cellular telephone makes it possible to broadcast live "remotest" Restrictions apply: as a non-commercial station, we are limited by law from broadcasting from commercial establishments. The cost by the minute, also limits the number and length of remotes per year. Never call any number except WGFR's. If the monthly bill shows any calls to other members, they will be billed to the individual who checked out the instrument.

## Graduation Videos

Broadcasters make real impact at A.C.C.! They videotape and sell each ACC graduation (May, August, and January). Traditionally, profits are donated to the A.C.C. Foundation to help with scholarships, etc. We need three people for each: one to take orders and collect money, and three people to direct and run the cameras. You'll assist in creating a tape which people will watch again decades in the future!

The Graduation Videos began during the 1992–1993 academic year.

## New Ideas

What are your ideas? Finish this column!

## Manner of Dress

Dress appropriately for the event. A wedding reception requires different clothing (men probably should wear ties) than a high school Road Show. You are exposed when videotaping a graduation, so caps, jeans and t-shirts with cute sayings don't make it. You also need to look neat when doing remote broadcasts, as you represent WGFR, ACC, and the College to the public.

Don't smoke, and don't chew gum.

## Thank Everyone

- Thank your client who pays for the Road Show.

- Thank your hosts who allow you to videotape.

- Thank anyone interviewed.

- Thank the people who supply tables and chairs, or extension cords, or anything else.

# Adirondack Broadcast Association Constitution

Article I Name

The name of this student organization shall be the Adirondack Broadcast Association, hereafter called the "ABA."

Article II Purpose

The purpose of the ABA shall be to raise funds for and otherwise contribute to non-profit causes, using the members' skills in telecommunications wherever possible, and also to assist new broadcast students in adjusting to ACC college life.

Article III Membership

> 3.1 Student Membership
>> Membership shall be open to any ACC student who has paid the activity fee.
>
> 3.2 Faculty Membership
>> The ABA shall grant non-voting membership to any faculty member.

Article IV Officers

> 4.1 Composition
>
> 4.1.1 There shall be four officers consisting of: President, Vice-President, Secretary, and Treasurer.
>
> 4.1.2 There shall be additional officers as the ABA deems necessary.
>
> 4.2 Time of Election
>> The election of officers shall take place during the month of April each academic year.
>
> 4.2.1 All elected officers m maintain a QPA and CQPA of least 2.0 at the end of each semester
>
> 4.3 Term of Office.
>
> 4.3.1 Each officer shall serve one year or until his/her success is elected.
>
> 4.3.2 Newly elected officers shall take office on September first of each academic year.
>
> 4.4 Method of Election
>
> 4.4.1 Each officer shall be elected by voice or secret ballot by the members present on the day election.
>
> 4.4.2 A simple majority of the ballots shall determine the winner. No proxy voting.
>
> 4.5 Number of Offices Held.
>> No person shall hold more than one office at the same time.
>
> 4.6 Nominations for Presented Slate
>
> 4.6.1 The President with the advice and consent of the other officers shall prepare and distribute a slate of nominees to the

general membership no later than five school days prior to the date set aside to vote.

4.6.2 Any person nominated must be present at the meeting at which nominations take place or acknowledge in writing that he/she is willing to be nominated for an office.

4.7 Nominations from the Floor.

Additional nominations may be made from the floor. Candidates shall not be nominated in absentia without the individuals' written consent.

4.8 Resignation and/or Removal from Office.

4.8.1 Upon the resignation of the Vice-President, Secretary, and/or

Treasurer, the office(s) shall be filled at the next regularly scheduled or special meeting. The newly elected officer(s) shall take office upon election and serve out the remaining term.

4.8.2 At any regularly scheduled meeting, the ABA's membership may by a two-thirds vote of the members present, remove any officer from office providing such officer has been duly notified no later than five school days before such action is to take place. Proxy votes shall not be counted toward the two-thirds vote.

Article V

5.1 General Duties.

The officers shall perform the duties prescribed by this document, as directed by the ABA or all other such duties as are customarily attributed to the office.

5.2 The duties of the President shall be to:

5.2.1 Preside over all regularly scheduled and special meetings of the ABA.

5.2.2 Attend all Inter-Club Council meetings.

5.2.3 Appoint all ad hoc committees with the advice and consent of the membership.

5.2.4 Attend the Student Senate's Budget Committee meeting along with the ABA Treasurer.

5.2.5 Prepare a slate of officers (see 4.6.1).

5.2.6 Prepare an agenda for all regularly scheduled and special meetings.

5.3 The duties of the Vice-President shall be to:

5.3.1 Preside over all regularly scheduled and special meetings in the absence of the President.

5.3.2 Assume the duties and responsibilities of the office of the President for the remainder of the President's term if:

5.3.2.1 The President is unable to function as the ABA's president for any reason, or

5.3.2.2 The President is impeached.

5.3.3 Serve at the behest of the President.

5.4 The duties of the Secretary shall be to:

5.4.1 Maintain an accurate record in the form of minutes of all deliberations and actions taken by the ABA at all regularly scheduled and special meetings.

5.4.2 Keep an accurate record of attendance at all regularly scheduled and special meetings.

5.4.3 Maintain a file of all previous minutes of the ABA.

5.4.4 Forward to the Director of Student Activities'* office an official list of newly elected officers and the faculty advisor(s) for the next academic year.

*Mr. Dale Solotruck, Student Center office

5.4.5 Forward to the Director of Student Activities' office any change in officers or faculty advisor(s).

5.4.6 Serve at the behest of the President.

5.5 The duties of the Treasurer shall be to:

5.5 .1 Maintain an accurate record of all monies in the form of income, disbursements, and current account balances.

5.5.2 Prepare a Treasurer's Report and present it to the membership at all regularly scheduled meetings.

5.5.3 Meet regularly with the Faculty-Student Association's office manager* to reconcile the ABA's financial records.

5.5.4 Prepare the ABA's annual budget, present it to the ABA for its approval, and then present it to the Student Senate's Budget Committee.

5.5.5 Maintain a file of all previous Treasurer's reports of the ABA.

5.5.6 Serve at the behest of the President.

Article VI

6.1 Frequency

6.1. 1 Regularly scheduled meetings of the ABA shall take place on the Adirondack Community College campus at least once a month during the academic year, except for January.

6.1 .2 Special meetings shall be called by the President after receiving a written request from five or more members in good standing. Such request shall contain the agenda for the special meeting and the President shall call the meeting no more than five school days after receipt of such request.

6.2 Faculty Advisor

No regularly scheduled or special meeting can be held without the faculty advisor in attendance.

Article VII Faculty Advisor

There shall be at least one faculty advisor elected by a two-thirds vote of the members present at the same time as the

ABA's annual election for its officers is held. A faculty advisor can be changed by a two-thirds vote of the members present at any regularly scheduled or special meeting. The Secretary of the ABA shall notify the Director of Student Activities'* office of such change.

* Mr. Dale Solotruck, Student Center office.

## Article V III Voting

### 8.1 Definition
Each member shall be entitled to one vote per issue.

### 8.2 Method
Voting shall be by voice or secret ballot.

### 8 .3 Secret ballot.
Any member may request a secret ballot on any issue.

### 8.4 Proxy
8.4.1 Proxy votes shall be in writing, signed and dated.

8.4.2 Proxy votes shall be allowed for voting purposes at any regularly scheduled or special meeting except as noted in this document.

8.4.3 No member shall cast more than one proxy vote per issue.

8.4.4 Members holding proxy votes shall inform the presiding officer before voting begins on the issue in question.

## Article IX Quorum

A quorum for a meeting shall consist of no less than six voting members. Proxy votes shall not be counted toward the quorum.

## Article X

### Parliamentary authority
Parliamentary authority for all regularly scheduled or special meetings shall be Robert's Rules of Order (revised edition).

## Article XI Amendments and Ratification

### 11. 1 Amendments
Amendments to this document shall be proposed in writing at any regularly scheduled or special meeting. Voting on such amendments shall take place at the next regularly scheduled or special meeting beyond five school days of the meeting at which the amendments were proposed.

### 11.2 Ratification
Ratification of the amendments shall take effect upon a two-thirds vote of the membership present.

## Article XII

In the case of conflicts, this Constitution shall be subordinate to the Constitution of the Adirondack Community

College Student Association anal or the published policies of the College's Board of Trustees.

## Notes on ABA Functions

Halloween Trick or Treat Traditionally ABA hands out small toys rather than nutritionally unwise candy to youngsters. These must be ordered many weeks in advance. In 1990 ABA started Trick or Treat outside Cue & Curtain's long-established "Haunted House." This led to the all-club Trick or Treat in 1993. In the lounge clubs fabricate booths (ABA won lst prize in 1993). Rules require recycled materials and little or no expense.

Angel Tree All clubs buy Christmas toys for disadvantaged children.

ABA vs. Lady Mountaineers This basketball game which began in 1993 raises money for Prospect Child & Family Center. See Professor Darla Belevich.

SuperJam The annual rock concert, missed in '94, traditionally donates all profits to Prospect Child & Family Center. Years ago this exceeded $2000.

NACB Conference Typically the ABA budget includes funds for a small part of each person's trip to Providence.

Graduation ABA videotapes all Commencements for sale to graduates, with profits going to the ACC Foundation.

Underwriting for WGFR See the Underwriting pages in this Manual.

Recognizing that broadcast-related Club activities including Road Shows, Underwriting for WGFR, and videotaping of events such as sports, commencement ceremonies and the like publicly represent Adirondack Community College, such activities are not subject to club vote and/or appointment by Club officers but are under the direct supervision of the Club Advisor or another employee of ACC with approximate experience and knowledge in the radio-TV discipline. Students retain the right to vote on disposition of any monies earned by these activities. Adopted by Student Senate Dec. 3' 97.

*Because of past inequities, it shall be ABA policy that any expenses incurred by students in conducting charitable and similar ABA functions shall be borne entirely by those individuals.*

## Attending the NACB Conference in Providence

### Missing Classes

Your instructors have no authority to excuse you from any classes except their own. Attending this Conference is a valuable experience, but you are responsible for requesting the time from your other instructors end preparing any required makeup work.

## Reservations

Make two separate reservations, for the hotel, and for the Conference itself. You are responsible for making your reservations. You will need a credit card ready when you call for the hotel reservation. Tell the hotel that you are attending the NACB Conference to get the special rate! You will need the credit card to check in at the hotel!

The special rate was $107 per room per night in 1994, with four people in a room (slightly less with fewer people to a room), so it comes to $26.75 per night per person—really a bargain for a luxurious downtown big-city hotel.

Take money for gas, parking, incidentals, and food.

## How to Get There

It's about 217 miles from downtown Glens Falls to Providence. With stops, allow 5 hours.

1.  Take the Northway, I-87, south to Exit 7.

2.  At Exit 7, take State Highway 7 east toward Troy. 3. Take I-787 south to I-90.

4.  Take I-90 east (toward Boston).

5.  Continue on I-90—it becomes the Massachusetts Turnpike.

4.  Take Exit 10. Go south a few miles to US Highway 20. It's confusing . . . watch signs for 20 and/or 146.

5.  Go east on US Highway 20 several miles to State Highway 146. Watch the signs carefully here.

6.  Take State Highway 146 south to Providence. 7. Take Exit 22 in downtown Providence.

8. On the exit ramp bear left, following the sign for downtown Providence. As the ramp merges with the other ramps, stay in the center lane and turn right at the first light.

To reach the Omni Biltmore hotel, proceed through the next light (Exchange Street), bearing slightly left. Pull in to the Omni Hotel on the right side of the street.

To reach the Convention. Center, turn right at the Exchange Street light. This takes you directly into the parking building next to the Convention Center. The NACB Conference is at the Convention Center, not the Civic Center.

## Parking

It costs to park your car. Plan on this! Also tip the hotel valet a dollar for parking your car and also returning it.

The Conference takes place at the Rhode Island Convention Center, just a block from the Grand Heritage (former Omni Biltmore) Hotel. Check in at Conference Headquarters. If you have pre-registered, they will have everything waiting for you.

The NACB is primarily a student organization. In fact, NACB is the largest student media organization in the world. Its staff is young, mostly in their twenties. They will help you with any and all questions.

## Money

Your registration fee entitles you to all events at the NACB Conference. You are responsible for your meals, entertainment, transportation, accommodations, and miscellaneous expenses. Remember that as this is not an official college event and that you are traveling as a responsible adult. Your teacher cannot loan any money.

## Cautions and Warnings

You are over 18 and adults, and you are on your own. Your teacher(s) may attend the Conference but in no way in a chaperone mode.

Adirondack Community College maintains institutional membership in the National Association of College Broadcasters, making it possible for its students to attend at the "member" rate, and the Student Senate typically awards funding (within the Adirondack Broadcast Association budget) to pay a percentage of individual costs. But the College and the Faculty-Student Association do not sponsor such attendance, and carry no responsibility whatsoever for students who do attend.

Providence claims to be one of the safest cities in the U.S., but it is a large city. Go out only in groups. Stay out of doubtful areas, especially at night. In 1990, two ACC students were mugged. If someone knocks on your hotel door, be sure it's someone you know before opening!

You are strongly urged, as a matter of adult responsibility, to advise your family of your itinerary and location of accommodations.

## About Providence

Check out the Brown University area for interesting restaurants, open late, and excellent record shops.

Note the highly visible State Capitol building. Its dome is larger than the U.S. Capitol in Washington, and is said to be the second largest in the world (next to the Vatican in Rome).

Weather may be mild, but is more often cold and windy in November. Take a coat! You're near the ocean!

## Play-by-Play Basketball Broadcasts

### Preliminary Arrangements

1. Be sure there are enough people interested in and capable of doing play-by-play!

2. You also need announcers willing to ride the games in the Control Room.

3. Sports Director acquires a copy of the entire basketball schedule and delivers it to station management.

4. For each game, be sure a log is prepared allowing space for proper entries.

5. Request a table and chairs for the proper location (southeast corner of gym) from the office of the Director of Facilities, Warren Hall.

### Things to Take from WGFR

1. Mixer (portable board).

2. AC extension cord.

3. Microphones.

4. Cable to connect Mixer to wall connector in gym.

5. For away games, cellular phone, antenna, power supply, and mixer-to-phone patch cable

### Play-by-Play People Provide:

1. Radio tuned to WGFR for cues.

2. Paper and pencil for stats, etc.

### Connections

1. Locate wall connector on south wall of gym near the east rear door. This is a type-F screw-on connector as used for VHS video.

2. Connect firmly the two bares wires on the cable to the binding posts on the rear of the Mixer. Then screw the other end onto the wall connector.

3.  Turn on power and turn up microphone to pick up background noise and telephone WGFR to ascertain that audio is heard in the Remote Cue position.

## Before Game Begins

1.  Control Room places Remote in Cue to hear gym.

2.  People in gym talk to Control Room through microphone. For example, perhaps the game is scheduled to begin at 8:00, and appears to be on time. People in gym want to go on air at 7:55. It's 7:53. Tell Control Room, "Send it to us two minutes from now."

3.  Now listen on the radio for cue.

4.  At 7:55, Control Room operator says, "We now take you live to the ACC gym," and turns up the Remote.

## After Game is Underway

1.  Control Room monitors at all times. Whenever gym people want to send it back, say so, or use agreed upon cue. You may return it for 30 seconds or a minute for PSAs, or five or ten minutes at half-time for music.

2.  White playing music, etc. Control Room always throws Remote switch into Cue to hear the gym.

## Riding Gain

1.  Set your Mixer levels correctly! During exciting action your voices rise. Set pots for your loudest, even though normal talking barely moves the VU meter. Otherwise you'll distort when you get loud!

2.  Be sure an mics are set to the same level! If one voice is consistently much louder than another, people won't listen.

3.  Work closely to the mics! The gym is noisy, even with a small crowd, and they can drown you out.

4.  The play-by-play people haven't time to ride gain when the game becomes exciting. The Control Room operator does this.

5.  If the gym people have problems (such as unmatched mic levels), tell them so over the air during time outs, etc.

### Station Identification

1. The Control Room operator is responsible for legal ID as close to the hour as practicable. If no Time Out comes, quickly fade down the Remote pot say say, "WGFR, Glens Falls."

2. Be sure to log the times of the IDs, as well as the time the game begin and the time it ends.

3. Write in any changes (if the game is canceled, etc.) Remember that log corrections require a single line strikeout, and initials of the person who made the correction.

# Responsible Concert Planning

### Written, signed agreements with each band, specifying:

A. Date of performance.

B. Amount of and name to whom check is to be made out.

D. Clear statement that the specified amount covers everything and that there will no no additional charges whatsoever.

D. Statement agreeing to any promotion and advertising planned, or specifying what is or is not acceptable.

E. Statement detailing type and quantity of food and beverages expected.

F. Acknowledgment of understanding that state law prohibits alcoholic beverages on campuses.

G. In case of performance dates scheduled in or near winter, an escape clause allowing cancellation because of inclement weather by the A.B.A. up to a specified time, such as 12:00 noon the day of the event.

H. Some sort of understanding that all arrangements on this least are near-complete before a firm "go-ahead" is set. It may be desirable to specify a time limit.

I. Length of set and approximate starting time (not that this determines the sequence of bands, which must be satisfactory to all participants).

J. A statement that the light and sound services retained will be fully satisfactory to the bands.

### Written, signed agreements for lights and sound, specifying:

A. Points A through H, above.

B. Preparation and scheduling of any radio advertising.

C. Preparation of legally-acceptable promotions for use on WGFR, with care to stay within FCC guidelines.

D. Delegating students to distribute posters and handbills in a proper and non-littering manner to record stores, schools, etc. Be sure permission to put up posters is obtained, especially at the all-important high schools. Point out that it's for charity. (Some students tend to become over-enthusiastic about their ability to distribute such promotional material, giving up after one or two refusals from store proprietors and discarding the remainder or distributing them improperly. It is necessary to delegate only reliable individuals for this responsibility).

Remember, there is no such thing as over promotion.

### Bands

*Students purporting to represent A.B.A. have, in past semesters, promised or implied participation in rock concert events to bands without the knowledge of the student organization as a whole and/or the advisor. Because these bands have been highly displeased by being "led on, " some mechanism for avoiding this problem needs to be devised*

"Tons and tons of bands will want to play Superjam" writes Alex Hyatt of Whitehall, ABA President and Producer of the vastly successful 1991 concert. "I had a problem when people within the ABA promised to get friends' bands a spot on Superjam. Another problem came up when I suggested that we hold an open vote on what bands should appear. People in various bands showed up like lobbyists. Advice: don 't disclose when the decision will be made, or by whom. I got several phone calls at home from bands who felt they had been cheated out of performing, but that wasn't the case. They were just bad bands. Limit backstage passes top a specific number per band if you want to keep a handle on security."

## Broadcasting from Remote Locations

When doing a "remote" from a commercial location, you may identify that location, but you may not urge people to come and see you there! There may be no call to action. You can say, "We're at Smith's Store in the Mall." You may not say, "Come on down to see us at Smith's Store in the Mall." Say it informationally, in a conversational tone, and at reasonable intervals, not constantly.

When doing a remote for charitable purposes, first talk with your subject before calling into the station. Find out what your subject plans to talk about, so that you can ask intelligent questions. Caution your interviewee to mention the place of business only in passing, not in a promotional sense. And plan to keep it short and concise. Two or three minutes are long enough!

We may not broadcast information about even charitable fundraising if it is tied to a commercial venture. For example, if the promotion is that Smith's Store will donate $1 to the charity for every $5 purchase, we cannot mention that on the air.

A non-commercial station such as WGFR must emphasize informational matters rather than actual fundraising. It is legal within narrow limits to urge people to donate to the Hyper model Viral Disease Research Fund, but it is much better to broadcast PSAs and short live interviews from experts about the dangers of Hyper model Viral Disease.

Caution: people you talk to on the air often want to be helpful, and may try to emphasize the business. For example, "Oh, yes, I come to Smith 's to shop as much at Smith 's as I can . . . " As a non-commercial station, we cannot broadcast such comments. Change the subject without being too obvious!

Also, see the information on legal underwriting.

## How To Do It

1. After preparing in advance (see the left column), telephone WGFR. Tell the operator on duty that you are ready to go on the air.

2. Wait for a cue. Either listen on the telephone or on a radio receiver for your operator to announce a cue such as, "We now take you live to the Mall."

3. Be sure that everyone speaks closely into the telephone mouth-piece or microphone.

4. Keep it short and concise! While there is no limit on the amount of air time which a non-commercial station may use for fund-raising for itself, such a station may not interrupt regular programming for fund-raising for others no matter how worthy the cause. Therefore any remote broadcasts must be short, and informational (news-like) in nature.

5. Give a clear end cue, such as, "We now return you to the WGFR Control Room".

When broadcasting from a remote location remember that you are at work, and also representing your station and your college. Act appropriately (no goofing off), and be neatly dressed.

# Broadcast Graduates (All Radio-TV Broadcasting AAS unless otherwise stated)

May 1985
(All Certificates)
John Antonak
Michelle Bakken
Dan Ladd
Dan Miner
Mark Mulholland
May 1986
(All Certificates)
John Amrock
Tim Celeste
Brian Delaney
Gary Guilfoyle
Anne Healy
May 1987
(Certificates except note)
Charles Clough
Kathy Dolezsar
Joseph Donahue
Brad Wood
Steve Cole (Degree)
May 1988
Nika DeSautels
Al Garrow
Susan Weber
May 1989
Rebecca Howe AA
Rob Robichaud AS
Bob DeSautels
Chris Dodd
Ken Mark
William Moore
Erin Shurner
Amanda Delczeg (Media)
Dave Vandenburgh (Cert.)
Summer 1989
Mark Mattus
Winter 1990
Kelly Pierce Certificate
May 1990
Mike Basso
Mark Bradway
Rob Gleason
Steve Gonzalez
Adam Klos

Dave McWhorter
Peter Tobey
Greg Truesdale
Christina Yattaw AA
Summer 1990
Ken Harrington
Gina Totzeck AA
Winter 1991
Derek Buchal
Brett Lange
Carla Garuccio
Sue Bartow (Certificate)
May 1991
Scott Brumley
Tom Case
Meghan Kiernan
Mike Maiorella
Jennifer Ollivett
Scott Reside
Brett Scoville
Darcy Vigneault
Summer 1991
Tony DiDonna
Dan Lewza
Charles Luce
Jeff Mercer
Winter 1992
David Jones
Rob Pope
Matt Powers
Mark Ramsey
Michael Chimiak (Media)
May 1992
Richard Cavak
Tom Chapla
Craig Elsworth
Alexander Hyatt
Ronald Rushia
Winter 1993
Todd Campbell
Mike Fulmer
Pete Christian
Tim Welch
Cindy Watrobski
May 1993

Don Sexton
John Barcomb
Ken Cornstock
Jason Wentworth
Tim Hamer
Josh Greene
Christine Mohr
Eric Lansing
Becky Denaker
Donna Grant
John Culver (Cert.)
August 1993
Frank Milam
Cathy Sabella
January 1994
Diane Creser(Cert.)
May 1994
Gordon McGrath
Jeff Scellen
Matt Twardy
Jaime VanDoren
Summer 1994
Jim Kelsey
David Korol (Certificate)
Stephanie Roe
Winter 1995
Jennifer Brown
Tom McNamara
David Mishler
May 1995
Kristen Frazer
Charles Galuski
Maiken Holmes (Math &
    Science major)*
*Not a broadcast major but
    active in broadcasting
August 1995
(Yon-HeeKim*) *Native of
    Korea, transferred before
    graduating to San
    Francisco State
January 1996
Megan Rocque
May 1996
Adam DeVoe

Karen Gregory
Larry Peleggi
(Robbie Wright*)
　　*Transferred just before
　　finishing degree
January 1997
Sebren Carmichael
Justin Chabot

Jeff Frodyma
May 1997
Darlene Erck
Mark Fitzgerald
Brian Gross
Shawn Hilleboe
Adarn Srnatko
January 1998

Wendy Archard
Rob Lutz
Kim Rauch
Ryan Turnbaugh
Media Arts major, extremely
　　active in WGFR as MD
Kris O'Connor transferred
　　before graduating

## Interesting Books for Broadcasters

If you're really interested in broadcasting, our library has some great up-to-date books.

Broadcast Voice Handbook: How to Polish Your On-Air Delivery, by Ann S. Utterback Call number: PN 1991.8 A6 U87 1995

Making Television Programs: A Professional Approach, by Richard Breyer Call number: PN 1992.75 B735 1991

Writing for Video, by Gene Bjerke Call number: PN 1992.7.B54 1997

Video Lighting and Special Effects, by James R. Caruso Call number: PN TK 6643 C37 1991

Any book by Eric Barnouw, leading historian of American television. Now retired, he lives nearby in Fair Vermont.

## COM 172 Introduction to Telecommunications Sound Blaster Editing Assignment

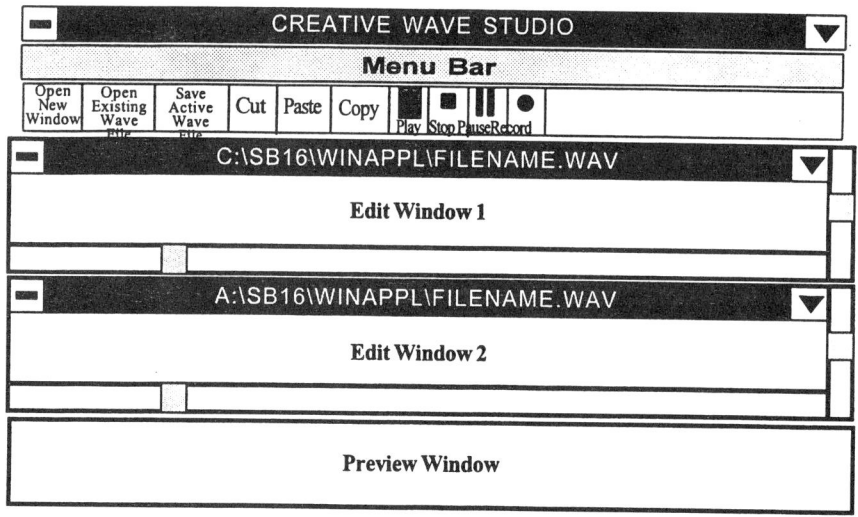

## Digital Editing Assignment

Here are Pesha's answers: "I'm glad to be here." "Thirty-six years." George Washington, Albert Einstein, and Arnold Schwarzenegger" "My wife says I did." "Why not?"

- Write your questions, making sure that the total "interview" time will be within 60 seconds.

- Open SoundBlaster and make sure Recording Settings at Mono, CCITT A-Law, and 11 kHz sample rate.

- Record your questions in Sound LE, making sure that your level is correct to avoid distortion.

- Open Wave Studio.

- Click on File, Open and from the C: drive the file name pesha.wav

- Play back to hear it.

- Click on File, Open and from the A: drive the file with your questions.

- Under Window click on Tile Horizontally. (Later you may prefer Tile Vertically).

- Play your file. You can play any file, determined by clicking on the Title Bar of the desired Edit Window. This file also shows in the Preview Window along the bottom, with the length in seconds and the recording characteristics showing in the Status Bar just below the Preview Window.

- Now insert the questions before the answers using Edit Copy and Edit Paste.

- When through play it back. If complete, Save it to your disk in the A: drive with an appropriate file name.

## Problems

- If the level of your file is too high or too low, alter them using Special and Amplify Volume.

- If the Recording Settings of your questions differ, use Special and Convert •Format. Can't see all the wave form? Use the vertical slider on the right side of the Edit Window.

- How much recording time remains? Under Options select either Display in Bytes or Display in Milliseconds. This indicates how much you can fit on your disk. For example, a High Density disk can accept about 1,400,000 bytes or 1.4 Mb. A millisecond is a thousandth of a second; therefore 32800 ms equals 32.8 seconds, etc.

- Experiment and have fun!

## How to Sign On

1. After having the door unlocked, turn on both wall switches beside the door.

2. Find the log dated for today (or date a blank log) and sign yourself on it.

3. Power switches on the board, CDs, cassette deck, and DAT should be on. If not, turn them on.

4. Power up the computer to SoundBlaster.

5. Prepare the recorded sign-on and your first recordings.

6. Check the EAS Monitor (see box below).

7. Assuming that normal programming is heard from the monitored station, turn on the transmitter.

8. Write Transmitter On time in the place provided on the log, and sign this entry.

9. Play the recorded sign on message and log this time.

## Ascertaining Correct Operation of the Emergency Action system

Broadcast stations many not operate without a correctly functioning EAS Monitor. Every time you sign on, you must do this first:

1. Listen to the primary station being monitored by depressing the SPKR button.

2. Listen for normal programming. If programming is obviously normal, then

3. Press RESET.

This must be done before sign on. The FCC fines stations for inoperative EAS monitoring.

If normal programming from the primary station being monitored is not heard, do not sign on.

## How to Sign Off

1. Play the recorded sign off.

2. Immediately when it finishes, turn off the transmitter switch.

3. Write Transmitter Off time in the place provided on the log, and sign this entry.

4. Sign yourself off the log.

5. Close CD drawers, and set the turntable speed controls to neutral.

6. Put away all your recordings and other stuff, and leave the control room neat.

7. Shut down the computer properly.

8. When leaving, turn off wall switches beside the door. Power switches on the board CD players, DAT, and tape deck should remain on.

9. If it's the end of the day, place the log in the designated place.

## Emergency Alert System

If the EAS Monitor sounds, attend to it at once regardless of the effect on programming. Listen carefully. Read the printout. If it is a real national emergency, WGFR must sign off the air. Otherwise staple the printout to the log.

In Case an FCC Inspector Visits

1. Ask for identification.

2. Try to contact the instructors, or Dean Arends in Warren Hall.

3. The "Public File" is in Dean Arends' office.

4. The logs, operator permits, and FCC Rules are filed in the instructors' office.

## Telephone Calls

Never accept collect calls. Any such calls must be paid for by the operator signed on duty. Records in the Registrar's Office will be impounded if collect calls are not paid for.

Never allow live telephone calls on the air. ACC liability insurance forbids it. Anyone doing so will be dismissed from WGFR permanently.

# WGFR
## 92.7 FM

### Glens Falls, NY
### Program & Operating Log

Day _____
Date _____1998

Scheduled programs approximately 2 minutes in length. All other times consist of music except as noted. Each Public Service Announcement (PSA) nominally 30 seconds. Transmitter is operated by remote control from main studio located at Adirondack Community College, Washington Hall, 640 Bay Road, Queensbury, New York.

Transmitter On time: _____Signature _____
Transmitter On time: _____Signature _____
Transmitter On time: _____Signature _____

Operators signed on and off below are on transmitter duty and program log duty. Each operator signs on when going on duty, and again, separately, when going off duty, attesting that this log accurately represents actual programming and transmitter operation during his/her period of duty.

## Check EAS for normal operation before every sign on!

_____time on _____    _____ time off _____

_____time on _____    _____ time off _____

_____time on _____    _____ time off _____

_____time on _____    _____ time off _____

_____time on _____    _____ time off _____

_____time on _____    _____ time off _____

_____time on _____    _____ time off _____

_____time on _____    _____ time off _____

_____time on _____    _____ time off _____

_____time on _____    _____ time off _____

_____time on _____    _____ time off _____

Transmitter Off time:_____Signature _____

Transmitter Off time:_____Signature _____

Transmitter Off time:_____Signature _____

Log reviewed at time: _____date: _____and found correct, unless so noted below, and signed below.

# Appendix E ▶ ▶

## FCC FORM 340

### Application for Construction Permit for
### Noncommercial Educational Broadcast Station

(Carefully read instructions before filing form) Return only form to FCC

| Section I—General Information | For Commission Use Only File No. |
|---|---|

| 1. Name of Applicant | Send notices and communications to the following person at the address below:<br><br>Name |
|---|---|
| Street Address or P.O. Box | Street Address or P.O. Box |
| City          State          Zip Code | City          State          Zip Code |
| Telephone Number *(include Area Code)* | Telephone Number *(include Area Code)* |

2.   This application is for:      ☐ AM          ☐ FM          ☐ TV

| (a) Channel No. or Frequency | (b) Principal Community | City | State |
|---|---|---|---|

(c)   Check one of the following boxes:

☐     Application for NEW Station

☐ MAJOR change in licensed facilities; call sign: _ _ _ _ _ _ _ _ _ _ _ _ _ _ _ _ _ _

☐ MINOR change in licensed facilities; call sign: _ _ _ _ _ _ _ _ _ _ _ _ _ _ _ _ _ _

☐ MAJOR modification of construction permit; call sign: _ _ _ _ _ _ _ _ _ _ _ _ _ _

File No. of construction permit; call sign: _ _ _ _ _ _ _ _ _ _ _ _ _ _ _ _ _ _ _ _ _

☐ MINOR modification of construction permit; call sign: _ _ _ _ _ _ _ _ _ _ _ _ _ _

File No. of construction permit; call sign: _ _ _ _ _ _ _ _ _ _ _ _ _ _ _ _ _ _ _ _ _

☐ AMENDMENT to pending application: Application File Number: _ _ _ _ _ _ _ _ _

Note: It is not necessary to use this form to amend a previously filed application. Should you do so, however, please submit only Section I and those other portions of the form that contain the amended information.

3.   Is this application mutually exclusive with a renewal application?   ☐ Yes   ☐ No

If Yes, state:

| Call letters | Community of License | | |
|---|---|---|---|
|  |  | City | State |

243

## Section II—Legal Qualifications

Name of Applicant

1.  Applicant is: (check one box below)

    ☐ (a)  government or public educational agency, board or institution

    ☐ (b)  private nonprofit educational institution

    ☐ (c)  nonprofit educational corporation

    ☐ (d)  other (specify)

2.  For applicants 1(c) or (d), describe in an Exhibit the nature and educational purposes of the applicant.

    | Exhibit No. |
    | --- |

3.  For applicants 1(c) or 1(d) applying for a new noncommercial educational television station only, describe in an Exhibit how the applicant's officers, directors and members of its governing board are broadly representative of the educational, cultural and civic segments of the principal community to be served.

    | Exhibit No. |
    | --- |

4.  Describe in an Exhibit how the proposed station will be used, in accordance with 47 C.F.R. Section 73,503 or Section 73.621, for the advancement of an educational program.

    | Exhibit No. |
    | --- |

    ☐ Yes  ☐ No

5.  Is there are any provision contained in any by-laws, articles of incorporation, partners agreement, charter, statute or other document which would restrict the applicant in advancing an educational program or complying with any Commission rule, policy or provision of the Communications Act of 1934, as amended?

If Yes, provide particulars in an Exhibit.

| Exhibit No. |
| --- |

## Citizenship and Other Statutory Requirements

6.  (a)  Is the applicant in violation of the provisions of Section 310 of the Communications Act of 1934, as amended, relating to interests of aliens and foreign governments? (See Instruction B to Section II.)

    ☐ Yes  ☐ No

    (b)  Will any funds, credits or other financial assistance for the construction, purchase or operation of the station(s) be provided by aliens, foreign entities, domestic entities controlled by aliens, or their agents?

    ☐ Yes  ☐ No

If the answers to (b) above is Yes, attach an Exhibit giving full disclosure concerning this assistance.

| Exhibit No. |
| --- |

7.  Has an adverse finding been made or an adverse final action been taken by any court or administrative body as to the applicant, any party to this application, or any non-party equity owner in the applicant, in a civil or criminal proceeding brought under the provisions of any law related to the following: any felony; mass media related antitrust or unfair competition; fraudulent statements to another governmental unit; or discrimination?

    ☐ Yes  ☐ No

    If the answer is Yes, attach as an Exhibit a full disclosure concerning the persons and matters involved, including an identification of the court or administrative body and the proceeding (by dates and file numbers), and a description of the disposition of the matter. Where the requisite information has been earlier disclosed in connection with another application or as required by 47 C.F.R. Section 1.65, the applicant need only provide: (I) an identification of that previous submission by reference to the file number in the case of an application , the call letter of the station regarding which the application or Section 1.65 information was filed and the date of filing; and (ii) the disposition of the previously REST CUT OFF)

    | Exhibit No. |
    | --- |

## Parties to the Application

8.  Complete the following Table with respect to all parties to this application

    (**Note:** If the applicant considers that to furnish complete information would pose an unreasonable burden, it may request that the Commission waive the strict terms of this requirement with appropriate justification.

    **Instructions:** If applicant is a corporation or an unincorporated association with 50 or fewer stockholders, stock subscribers, holders of membership certificates or other ownership interest s, fill out all columns, giving the information requested as to all officers, directors and members of governing board. In addition, give the information as to all persons or entities who are the beneficial or record owners of or have the right to vote capital stock, membership ownership interests or are subscribers to such interest. If the applicant has more than 50 stockholders, stock subscribers or holders of membership certificates or other ownership interest s, furnish the information as to officers, directors, members of governing board, and all persons or entities who are the beneficial or record owners of or have the right to vote 1% or more of the capital stock, membership or ownership interest s. If applicant is a governmental or public educational agency, board or institution, fill out columns (a), (b), and (c) as to all members of the governing board and chief executive officers.

| Name and Residence Address(es) | Office Held | Director or Member of Governing Board<br>Yes    No | % of Ownership (O) or Voting Stock (VS) or Membership (M) |
|---|---|---|---|
| (a) | (b) | (c) | (d) |
| | | | |

9.  Does the applicant, or any party to the application, have a petition to migrate to the expanded band (1605–1705 (kHz) or a permit or license either in the existing band or expanded band that is held in combination with the AM facility proposed to be modified herein?     ☐ Yes  ☐ No

    If Yes, provide particulars as an Exhibit.

    ┌─────────────┐
    │ Exhibit No. │
    └─────────────┘

10. Does the applicant or any party to the application have, or have they had, any interest in:

    (a)  a broadcast station, or pending broadcast station application before the Commission?     ☐ Yes  ☐ No

    (b)  a broadcast application which has been dismissed with prejudice by the Commission?     ☐ Yes  ☐ No

    (c)  a broadcast application which has been denied by the Commission?     ☐ Yes  ☐ No

    (d)  a broadcast station, the license of which has been revoked?o     ☐ Yes  ☐ No

    (e)  a broadcast application in any pending or concluded Commission proceeding which left unresolved character issues against the applicant?     ☐ Yes  ☐ No

    If the answer to any of the questions in (a)–(e) above is Yes, state in an Exhibit the following information:

    ┌─────────────┐
    │ Exhibit No. │
    └─────────────┘

    (1)  Name of party having interest s;

    (2)  Nature of interest or connection, giving dates;

    (3)  Call letters of stations or file number of application or docket; and

    (4)  Location.

## Section III—Financial Qualifications

**Note:** If his application is for a change in an operating facility DO NOT fill out this Section.

1.  Is this application contingent upon receipt of a grant from the National Telecommunications and Information Administration?    ☐ Yes  ☐ No

2.  Is this application contingent upon receipt of a grant from a charitable organization, the approval of the budget of a school or university, or an appropriation from a state, county, municipality or other political subdivision?    ☐ Yes  ☐ No

**Note:** If either Questions 1 or 2 is answered "Yes," your application cannot be granted until all of the necessary funds are committed or appropriated. In the case of grants from the National Telecommunications and Information Administration, no further action on your part is required. If you rely on funds from a source specified in Quest ion 2, you must advise the F.C.C. when the funds are committed or appropriated. This should be accomplished by letter amendment to your application, in triplicate, signed in the same manner as the original application, and clearly identifying the application to be amended.

3.  The applicant certifies that sufficient net liquid assets are on hand or that sufficient funds are available from committed source to construct and operate the requested facilities for three months without revenue.    ☐ Yes  ☐ No

## Section IV—Program Service Statement

Attach as an Exhibit, a brief description, in narrative form, of the planned programming service relating to  the issues of public concerning facing the proposed service area.

| Exhibit No. |
| --- |

**Note:** No program service statement need be filed where the proposed station's programming would be wholly "instructional" as that type of programming is defined in the instructions to this Section.

# Appendix F ▶ ▶

## HELPFUL DOCUMENTS FROM LPB, INC.

### Suggestions for Starting a Noncommercial Educational Broadcast Station

#### So, What Kind of Unlicensed Station Can I Start?

*AM Carrier Current*
- Systems place AM signal on electrical lines of the campus
- Mono Signal
- Anywhere on AM band 530–1700 kHz*
- No FCC License needed
- Typically one package (Transmitter + Coupling Unit) per building
- Commercials permitted
- Typically maximum of $1150/building, generally much less
- Site survey recommended

*AM Vertical Antenna*
- Systems broadcast via vertical antenna, typically 20 ft in height
- Mono Signal
- Anywhere on AM band 530–1700 kHz*
- No FCC License needed
- Typically one package per campus, outside coverage only
- Commercials permitted
- Typically maximum of $2500/campus
- Site survey recommended, careful engineering Required

*FM Radiating Cable*
- Systems broadcast via radiating coaxial cable, installed inside buildings  on one floor
- Mono/Stereo Signal
- Anywhere on FM band 88–108Mhz*
  No FCC License needed
- Typically one package (Transmitter + Radiating Cable) per building
- Commercials permitted
- Typically maximum of $2700/building, generally much less
- Site survey recommended, post-installation testing Required

### FM Cable TV Broadcasting

- Systems place FM signals on campus or local Cable TV systems
- Mono/Stereo Signal
- Anywhere on FM band 88–108Mhz*
  No FCC License needed
- Typically one package (Transmitter) per campus
- Commercials permitted, subject to approval by CATV operator
- Typically maximum of $1110/campus
- Site survey not needed

*FCC Part 15 requires that these stations may not cause interference to any licensed operation.


Federal Communication Commission 10/28/1997
Audio Services Division


Mass Media Bureau—Federal Communications Commission
3rd Floor—1919 M Street NW—Washington, DC 20554

### Audio Services Division Telephone Numbers for Application Status

Assignment and Transfer Applications . . . . . . . . . . . . . . .202-418-2782

Radio Renewal Applications . . . . . . . . . . . . . . . . . . . . . .202-418-0186 or
. . . . . . . . . . . . . . . . . . . . . . . . . . . . . . . . . . . . . . . .1-800-671-2233 or
. . . . . . . . . . . . . . . . . . . . . . . . . . . . . . . . .by e-mail to radioren@fcc.gov


*Other applications:*

AM Service . . . . . . . . . . . . . . . . . . . . . . . . . . . . . .202-418-2795

FM Service . . . . . . . . . . . . . . . . . . . . . . . . . . . . . .202-418-2730

FM Translator Service . . . . . . . . . . . . . . . . . . . . . . . .202-418-2795

Silent Station Information . . . . . . . . . . . . . . . . . . . . . .202-418-2795

Engineering Database . . . . . . . . . . . . . . . . . . . . . . . .202-418-2795

Other Inquiries . . . . . . . . . . . . . . . . . . . . . . . . . . . .202-418-2782

Mass Media Bureau—Federal Communications Commission